THERE IS NOTHING NEW UNDER THE SUN

The story of Renewable Energy

Solomon D. Matios

All streams flow into the sea, yet the sea is never full.
To the place the streams come from,
there they return again.

What has been will be again,
what has been done will be done again;
there is nothing new under the sun.

Is there anything of which one can say,
"Look! This is something new"?
It was here already, long ago;
it was here before our time.

No one remembers the former generations,
and even those yet to come
will not be remembered
by those who follow them

The words of the Teacher, son of David, king in Jerusalem: Ecclesiastes Chapter 4-11

The cultural and scientific accomplishments of the past
have allowed humanity to build an advanced and prosperous world.
This book is dedicated to the wise men and women who
look back to move forward.

Contents

There is nothing new under the sun

New Energy Revolution...8

History of Fossil Fuels...15

Deteriorating Climate and Public Health 21

    Air pollution 21
    Land degradation 23
    Water pollution 27

Loss of Precious Spaces..29

    Theodore Roosevelt National Park......................................29
    Dakota Access Pipeline or Bakken Pipeline..........................31
    Arctic National Wildlife Refuge..32
    Worst oil spill disasters 35
        Exxon Valdez oil spill..35
        BP Deepwater Horizon oil spill.....................................35
        The Persian Gulf War oil spill 36
        Port Louis, Mauritius...36
        Sri Lanka 36
        Russian Arctic North oil spill...37

National Security and Global Conflicts.......................................38

The price of Change...44

    Job Creation 44
        Electric power generation...46
        Energy efficiency 47
        Fuels..47
        Transmission, Distribution & Storage............................47
    Cost of Electricity 47

Energy Efficiency 50

Building sector efficiency: Old techniques 51
- Passive solar heating 52
- Thermal mass 52
- Passive solar cooling 54

Building sector efficiency: New techniques 57
- Insulation 57
- Air leakage 59

Energy star program 61
- Modern appliances & accessories 61
- Energy star and real estate 61
- LED lights 63

Hydro Energy 65

Hydro energy types 67
- Impoundment 68
- Pumped storage 74
- Diversion (run-of-the-river) 76

Sizes 80

The nuts and bolts of hydropower 80
- Impulse turbine 81
- Reaction turbine 82

Advantages of hydropower 85

Wind Energy 86

Understanding wind energy 90
- Towers 92
- Blades 94
- Nacelle 96
  - Yaw System 97
  - Mechanical drive train 99
  - Electrical systems and cabinets 99

Wind turbine types 100
Site allocation 101
Wind farms 102
- Onshore wind farms 102
- Offshore wind farms 104

Small scale wind power 106
    Horizontal axis wind turbine (HAWT) 107
    Vertical axis wind turbine (VAWT) 107

Active Solar Energy 109

Types of solar panels 113
    Silicon solar panels 114
    Thin film solar panels 115
    Bifacial solar panels 116
    III-V solar panels 116
    Next generation solar panels 117
Additional hardware 117
    Inverters 117
    Charge controllers 118
    Batteries 119
Solar power installations 119
    Residential scale solar power 119
    Commercial scale solar power 119
    Community scale solar power 120
    Utility scale solar power 120
Solar thermal (heat) energy 120
    Small scale solar thermal energy 121
    Large or utility scale solar thermal energy 121
        Linear concentrator systems 122
        Dish/Engine systems 123
        Power Tower systems 123

Geothermal Energy 125

Source of geothermal energy 126
Types of extraction 129
Heating and cooling 129
Electricity 130
    Dry steam power plant1 130
    Flash steam power plant 130
    Binary cycle power plant 130
    Enhanced geothermal systems (EGS) 131
    The Geysers Geothermal Complex 132

Renewability and sustainability 134
Geothermal energy worries 134
Benefits of geothermal energy 135

Bioenergy 138

BIOPOWER for power utilities 138
BIOFUELS for the transportation industry 139
    Bioethanol 140
    Biodiesel 140
    Green diesel 141
    Straight vegetable oil 141
BIOPRODUCTS for the manufacturing industry 141
Investment and jobs 141
Bioenergy and the environment 142

Conclusion 143

References 145
Photo credits 147
Case Studies

Windmill for grinding grain 11
Vertical blades 11
Windmill for pumping water 12
Industrial Revolution 20
Methods of extracting oil 25-26
Mining Toxins 28
Resource conflict in the DRC 42-43
Ancient Wisdom (From Khartoum, Sudan to Colorado, USA) 57
Eritrean man-made lakes 69
Lake Mead and Hoover Dam, Arizona/Nevada, USA 70-71
Bonneville Dam, Oregon, USA 72-73
Raccoon Mountain pumped storage, Tennessee, USA 75
Chief Joseph Dam, Washington, USA 78
Willamette Falls Dam, Oregon, USA 79
The Danakil Depression, Eritrea 136

New Energy Revolution

> All streams flow into the sea, yet the sea is never full.
> To the place the streams come from,
> there they return again.

Human thoughts and behavior are similar to the flowing streams in many ways. Ideas come and go, only to be re-introduced by future generations. We allow ourselves to forget old ideas, only to re-introduce them as new. Today, we are in the midst of rejuvenating an old idea. An idea that has been gradually changing our energy consumption behavior. It is an idea that is about to change our climate for the better and improve our public health in the process. It is also an idea that is changing the way we do business locally and globally.

This idea may even be the beginning of a new revolution, a revolution driven by an ever-growing global population, deteriorating climate and worsening public health. According to the United Nations (UN), the global population has increased from about 1 billion to over 7 billion in just 200 years. To support this massive population increase, global economies are consuming an unsustainable amount of energy to power and heat homes, businesses and industries. Energy consumption for the transportation industry, agro-industry and the pharmaceutical industry, to name a few, are all at a record level. It is no wonder energy (mainly fossil fuels) consumption and its impact on climate change has dominated the airwaves for the past decade, often pinning public opinion against one another.

There is plenty of evidence that proves expanding demand and appetite for economic growth are leading to increased waste products that negatively impact the climate and public health. Multiple world organizations, such as the United Nations (UN) and World Health Organization (WHO), as well as domestic entities such as the United States Center for Disease Control (CDC), conclude that changing weather patterns, rising global temperature and increased CO^2 emissions are factual. These climate changes contribute to public health deterioration, leading to critical health concerns such as heat stress, asthma, and malnutrition.

In addition to climate change and public health deterioration, the pressure to support growing populations and, thus, the increasing demand for energy is leading to rising global conflicts for access to limited resources. There is also an extreme pressure on our finite energy resources as we continue to exploit precious and reserved territories, such as national parks, for additional energy sources. For those reasons, the dominant energy sources of fossil fuels (oil, coal, natural gas) are scrutinized and are falling out of favor for more cleaner, possibly cheaper and renewable energy sources like solar, wind, hydro, geothermal and biofuel.

On the other hand, there are public disagreements on climate change research and models' accuracy and consistency. Many believe climate change is natural. Fears of changing weather patterns and increased pollution are politically motivated. Thus, any policy decision based on inconsistent and inaccurate research and climate change model may demonize the fossil fuels industry. Policies and regulations based on misinformation and misunderstanding will ultimately lead to harsh economic disasters. Factories and refineries will shut down, jobs will be lost, economic opportunities will be forfeited, and livelihoods will be destroyed.

Before we discuss the details of both arguments, it only makes sense to understand what the alternative to fossil fuels is. The alternative is renewable energy, which is interchangeably used with alternative energy or clean energy. For discussions in this book, "alternative energy" or "clean energy" refer to non-fossil sources of energy that do not emit greenhouse gases to the atmosphere. These energy forms include sources from nuclear, solar, wind, biofuel, hydro, and geothermal, to name a few. On the other hand, renewable energy means any clean energy source that is infinite or can be replenished in a short amount of time, meaning all sources of energy, except fossil fuels and nuclear energy. Nuclear energy is not renewable because it relies on non-replenishable, mined minerals such as uranium. Fossil fuels are also non-renewable because they take millions of years to replenish. The focus of this book is renewable energy, non-fossil and non-nuclear.

Renewable energy is an old concept that has been utilized for thousands of years in many parts of the world. For example, ancient Egyptians are among the earliest on record to use wind power to propel boats along the Nile River as far back as 5000 BC. Another wind-powered creation is the windmill, which was very common in many parts of the ancient world. A windmill is a structure with rotating blades or sails that converts a blowing wind into rotational energy used in grinding grain and pumping water. Another form of ancient energy source is the waterwheel, which is the oldest form of hydropower. Waterwheels use the kinetic energy of flowing water to spin attached paddles or buckets. The kinetic energy is then converted to rotational or mechanical energy and used for various purposes, including supplying running water for irrigation, grinding grain, water pumps and energy sources for industry such as in sawmills, textile mills and cast iron making. Of course, let's not forget the earliest form of biofuel, animal dung, dried up in the open field and used as fuel for cooking or heating.

Common types of Renewable energy

Although many forms of renewable energy have been used for centuries, modern renewable energy began its journey in the early to mid-1800s with few scientific discoveries. Some of the findings include the work of Edmond Becquerel, a French physicist, who in 1839, discovered the effect of light and how it interacts with electrolytic cells. In the years to follow, many prominent scientists advanced the work. Honorary mentions include Professor William Grylls Adams of King's College in London and his colleague Richard Evans Day. In 1876, they demonstrated how certain materials, such as selenium, can produce an electric current when exposed to light. Fast forward to 1905, famed physicist Albert Einstein published a paper explaining the 'photoelectric effect, which is the emission of electrons when light is shined upon certain materials.

A few years later, Albert Einstein would win a Nobel Prize for Physics in 1921 for his work in the photoelectric effect, which is the driving force behind the modern solar energy industry.

Windmill for grinding grain

This structure uses the power of the wind to spin the attached blades, which are connected to a drive shaft. The gear at the end of the drive shaft is connected to another gear and an additional shaft. At the bottom of the structure is two millstones. One is stationary, while the other is attached to a spinning shaft. Grain is poured through a hole in the top millstone and ground by the rotational force of the millstones. Finally, the flour exits on the side of the stone into flour sacks or silos below.

Vertical blades

Earlier blades, like the one in Nashtifan, Iran, were vertical and were designed to catch the blowing wind easily. However, with time, blades were designed and installed horizontally and attached to a central drive shaft for better efficiency. Part of the design includes making the blades slightly angular (like airplane wings) to catch a blowing wind. In later chapters, we will elaborate on how this technique led to modern wind blade design shift from vertical to horizontal.

Windmill for pumping water

This structure also uses the power of the wind to spin attached blades. The rotating blades force an attached rod to pull up and down. The up and down motion of the rod helps pump underground water to a storage tank.

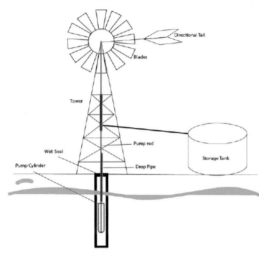

In the 18th century, American settlers in the Western Great Plains were challenged with a shortage of water for their personal needs, for watering their livestock and for growing crops despite the abundance of water deep underground. With the introduction of windmills from European migrants, they were able to resolve their water shortages by pumping water from great depths at a steady rate. Windmill pumps have the advantage of shifting into the prevailing winds. Thus, they are functional in both fast and slow currents. Most importantly, they require little maintenance. Soon, windmill pumps became a "must-have" for every homesteader, farmer and rancher in the American West.

windmill pump in a Texas farm in year 2021)

Not to be outshined, modern wind energy has a journey of its own. Besides the windmills and windpumps of yesteryears, wind turbines have been used for producing a small amount of electricity since the late 1800s. For example, Denmark, one of the most advanced countries in wind energy, started generating electricity from wind power as early as the 1890s. By 1908, Denmark had over 70 electricity-generating wind power systems.

By the 1920s, similar wind power systems were utilized among farmers and ranchers of the Western United States. Following the oil shortages of the 1970s, massive investment and government policies helped wind energy expand in the U.S. By the1980s, wind energy was integrated into the grid system of many utility companies to produce electricity for thousands of homes and businesses. Today, there are over 350,000 wind turbines in operation around the world generating electricity. According to the American Clean Power Association, the U.S. accounts for at least 60,000.

Other renewable sources, such as hydropower, have been around for a while, too. They work similarly to the ancient waterwheels. A large amount of water flows down a tunnel into the turbine and spins the turbine blades the same way flowing water spins waterwheel blades or buckets. The rotating turbine spins an attached shaft, which turns the electricity-producing generator. One of the prides of modern hydropower achievement is the Hoover Dam, located on the borders of Arizona and Nevada on the Colorado River. The Hoover Dam project began in 1931 and at its completion in 1935, was the largest hydroelectric facility in the world with an electric generating capacity of 2.08 gigawatts (GW). According to the U.S. Bureau of Reclamation, the Hoover dam generates 4 billion kWh of electricity annually for over a million homes in Nevada, Arizona, and California.

(left) Water wheel of yesteryears used to re-direct water for irrigation. (right) Hoover dam: a modern hydroelectric dam generating electricity for millions of homes annually.

Hydropower is so cheap and so reliable that many large projects have been built over the past few decades and continue to be built today. Some of the most recent projects include the "Three Gorges Dam" in China. When completed in 2012, it became the largest hydropower project in the world with an electricity production capacity of 22.5 GW, nearly ten times the capacity of the Hoover Dam. Thousands of

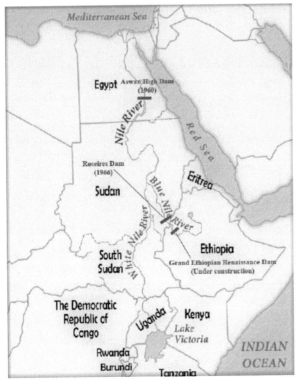

The newly constructed "Grand Ethiopian Renaissance Dam" on the Blue Nile

miles away, on another continent, the Ethiopian "Grand Dam" is one of the newest and still under construction with a planned capacity of 6.4 GW. When completed, the Ethiopian "Grand Dam" will be the largest in Africa, surpassing Egypt's Aswan Dam, which was completed in 1970 at a capacity of 2.2 GW.

With so much history and untapped potential, why did renewable energy remain undeveloped and overlooked for generations? Why didn't the renewable energy industry benefit from the scientific progress of the 1800s and the early 1900s? To answer these questions, we first need to understand the history of fossil fuels and how they have become the dominant energy source. After all, fossil fuels remain the chief energy source today and will probably stay as such for an extensive amount of time.

Fossil fuels are the result of plant and animal remains buried deep underground for millions of years. These decomposed remains have a high carbon and hydrogen content. Under extreme pressure and heat, the carbon and hydrogen molecules start to break apart and turn into peat and plankton. Over millions of years, the peat turns into coal while the plankton turns into natural gas and oil. Collectively, we know them as fossil fuels. Today, we mine the coal and drill for oil and natural gas to extract the stored carbon and hydrogen, which result in energy when burned.

Illustration of a simplified process of the origin of fossil fuels.

History of Fossil Fuels

Although the technologies to extract commercial level quantities are new, fossil fuels have been used in different civilizations worldwide for thousands of years. Historical references go back to biblical times and point out how the ancients used crude oil/tar for many purposes, including in construction and waterproofing boats. In Genesis 11:3, the bible describes how people in those days used bricks hardened with fire and tar (crude oil), found seeping on the surface of the ground, for mortar to build towers. The bible also makes another reference to tar in Genesis 6:14 and tells the story of God instructing Noah to build a large boat from cypress wood and waterproof it with tar, inside and out.

Around 4000 BC, the Sumerians of Mesopotamia region in modern-day Iraq used asphalt to glue their mosaic works on floors and walls. Ancient Egyptians used to it mummify their dead. Persians, Native Americans and Chinese all used crude oil for medicinal benefits as a skin treatment. In addition, crude oil was used for handles on arrows and knives and burn-weapons during warfare.

Floor and wall mosaic similar to ancient Sumerians art decor

Herodotus, an ancient Greek historian, also known as the father of European history, describes Ardericca, a military staging post, in the book "THE HISTORIES" in this way. "Ardericca is 210 stades from Susa, and it is also forty stades away from a well which is a source of three different products: bitumen, salt, and oil are all extracted from it." Herodotus was a well-traveled man and through his travel records, we get an in-depth historical reference to the times when he lived 485 BC to 425 BC.

Few centuries forward to the middle of the 19th century, people began to drill for crude oil purposefully. In 1859, Edwin L. Drake drilled the first commercially successful oil well in Titusville, Pennsylvania. The oil from these wells was refined to produce kerosene as fuel for fuel lamps. However, it wasn't until the invention of the first car with an internal combustion engine in 1885 by German engineer Carl Benz that crude oil would be processed for other than kerosene. Carl Benz's car was powered by a byproduct of kerosene called gasoline. Not far behind, Henry Ford launched the Model T in 1908, an affordable mass-market automobile in the U.S. The mass production of cars led to gasoline replacing kerosene as the most demanded crude oil's refined product. Additional oil discoveries in the Middle East propelled oil to become the most used energy source in the Western World in the early 20th century. In addition to oil demand for automobiles, both World War I & II created a demand for petroleum products as armies and navies expanded their fleets. Warships, tanks and military trucks consumed a large amount of fuel and helped oil demand soar worldwide.

The late 19th century witnessed the rise of the Standard Oil Company, founded by John D. Rockefeller in 1870, as the dominant oil company in the United States. Standard Oil Company controlled nearly two-thirds of the petroleum industry in the United States. Following the antitrust lawsuits in 1911, Standard Oil was split into many regional companies. Standard Oil of New Jersey/Exxon, Standard Oil of New York/Mobil, Standard Oil of California/Chevron would emerge as the dominant oil conglomerates in the years to come. Outside the United States, Royal Dutch Shell was established in 1907 following a merger between Shell Transport and Trading Company and Royal Dutch. Two years later, British Petroleum or BP emerged from the Anglo-Persian Oil Company.

Today about 70% of the transportation industry is fueled by petroleum, generating trillions of dollars for those conglomerates. Thus, any effort to demonize the petroleum industry and the conglomerates that control it, is unproductive and unfruitful. Their experience and financial power can play an instrumental role in the renewable

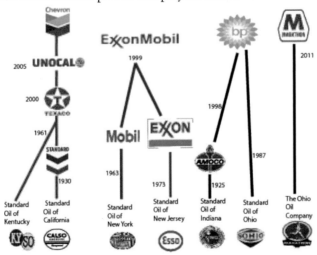

energy industry, as witnessed by the billions of dollars they have begun to invest in renewable energy.

If petroleum is the backbone of the transportation industry, coal was the backbone of the industrial revolution and later, a major source of electricity production. The Chinese were one of the earliest in recorded history to mine coal around 3000 BC. Marco Polo, on his trip to Asia in the late 1200s, described the use of coal in China; "It is a fact that all over the country of Cathay there is a kind of black stones existing in beds in the mountains, which they dig out and burn like firewood. If you supply the fire with them at night, and see that they are well kindled, you will find them still alight in the morning; and they make such capital fuel that no other is used throughout the country. It is true that they have plenty of wood also, but they do not burn it, because those stones burn better and cost less. Moreover, with the vast number of people, and the number of hot baths that they maintain--for everyone has such a bath at least three times a week, and in winter if possible every day, while every nobleman and man of wealth has a private bath for his own use--the wood would not suffice for the purpose." Centuries later in 15th and 16th century Europe, coal became an increasingly important fuel for heating homes with firebrick chimneys.

The industrial revolution that began in the mid-18th century elevated coal to the status of dominant energy source. Coal fueled the steam engine, which was the invention at the heart of the industrial revolution. Coal was available in abundance and it would soon replace water, which was the critical source of power before the Industrial Revolution, as the primary source of energy. The introduction of a more efficient steam engine in the mid-1770s made coal an essential energy source for industry and households. Coal-powered steam engines propelled the incredible powers of manufacturing and industrial activity and helped usher society into the modern era. Steam-powered machines made mass production possible. Coal also changed the way people traveled as steamships and steam-powered trains were burning coal to power their boilers.

In the Americas, coal has an early and humble beginning as well. Records show coal was an important fuel source for Native Americans in the 1300s. It was used for cooking, heating, and baking clay potteries. During the American Civil War of the 1860s, weapons factories started using coal in abundance. By the late 19th century, coke made from coal was the primary fuel for steel-making furnaces. During the same era, in the 1880s, utilities started using coal for electricity generation in the United States. Half a century later, by the early 1960s, coal had already become the dominant energy source of U.S. electricity generation.

Throughout centuries, coal consumption was steadily growing and outpacing the use of wood for energy. After all, coal was in abundance, easier to transport and gave off much more energy than wood. During their peak in the late 1990s to early 2000s, coal-powered utilities generated nearly 60% of all electricity in the United States. In addition to propelling the industrial revolution in Europe and helping power the electricity demand in the U.S., coal is also credited for powering other nations such as China into the modern era. At its peak in 2013, coal was powering about 78% of China's electricity generation and nearly 60% of its total energy consumption.

Coal/steam powered train from the early 1900s, still in operation in Asmara, Eritrea. Sign of workmanship of the industrial revolution era.

Today, the era of coal is on a decline. Coal is targeted for extinction due to its pollutant nature. After all, coal is the most pollutant of all fossil fuels. For that reason, coal-powered utilities are on the decline in the U.S. and worldwide. Replacing coal-powered utilities is another fossil fuel, natural gas. Recent advances in technology and know-how have allowed natural gas, which once was burned or flared as a waste byproduct of oil drilling, to become just as widely consumed as coal or oil. Today, natural gas is the most used fossil fuel in electricity production because it is much less pollutant. However, it still remains a non-renewable, unreplenishable source.

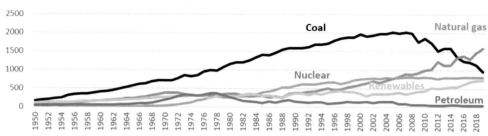

Fossil fuels are a significant energy source for everything our modern lives demand for residential, commercial and industrial consumption. They are also feed for many consumer goods, including plastics, medical equipment, paint and asphalt for roads and roof shingles. Fossil fuels are the primary enablers of the industrial revolution, which has allowed humanity to achieve so much in so little time: The world population has increased sevenfold in less than 300 years. People are living better, healthier and longer than their forefathers in most parts of the world. Technological advancement is at record highs. People are traveling from one end of the globe to the other at record speeds.

Unfortunately, increased fossil fuels consumption has led to concerns over climate change patterns, public health deterioration, loss of precious and protected spaces as well as rising global conflicts for resource dominance. For those reasons, there is negative public sentiment towards fossil fuels and a need to gradually shift towards a cleaner and renewable energy source.

In the following few chapters, we will try to gain a clear understanding of the different types of renewable energy sources and how we can benefit from them. We will also analyze the impact of renewable energy on global and local economies. We will evaluate if the renewable energy industry can provide as many economic opportunities and jobs as the fossil fuel industry has for hundreds of years. Most importantly, we will use market data and labor statistics to analyze how the industry competes on pricing. After all, most industrialized nations are spoiled with cheap energy prices from fossil fuels and nuclear energy-driven utilities.

One of the few remaining coal-fired power plant; Georgia, USA.

Industrial Revolution
The rise of fossil fuels

The industrial revolution, which began in Great Britain in the 1760s, is the evolution of economic advancement from agricultural to industrial.

There are four stages of the industrial revolution

The first industrial revolution: (1760-1830) The transformation of economies from agricultural to industrial. The era introduced mechanical production and textile mills using coal-powered steam engines and hydropower.

The second industrial revolution: (1870s-mid-1910s) This era revolved around the invention of the combustion engine, which was powered by gasoline. The discovery of electricity along with developments in machines, tools and computers, gave rise to the automation of factories. We were also introduced to steel, lighter metals, new alloys and chemically based products, such as plastics. The introduction of the telegraph jump started the communications industry, while the invention of planes and cars revitalized the transportation industry.

The third industrial revolution: (Late 20th century) is the era of the digital revolution and the era of nuclear energy. Research and development took off in the western world following the end of two world wars. Unfortunately, we witnessed the first use of the atomic bomb during this era. The electronics industry got its jump start during this time as well.

The fourth industrial revolution: (Current) Internet of Things, cloud technology, artificial intelligence and virtual reality are dominating this era. Regarding energy, we see a rapid shift from fossil fuels towards renewable energy such as solar, wind, hydropower, biomass and geothermal.

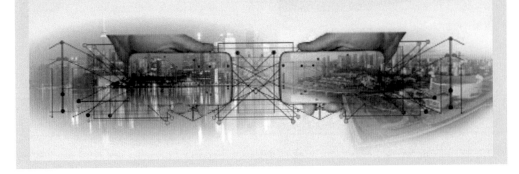

Deteriorating Climate and Public Health

In 1700, before the widespread use of fossil fuels, the world's population was about 670 million. By 2019 the world's population had reached over 7 billion. In the 20th century alone, the world's economy and energy consumption grew to a level never seen before. Many people around the world today enjoy the benefits of industrialization fueled by fossil fuels. For example, heavy duty machines do most of the labor intensive work, while intelligent devices help us perform better and quicker. Most of us benefit from improved health care and education. For example, child death rates have declined and life expectancy has increased in most parts of the world. Overall, mankind has achieved and benefited so much from the use of fossil fuels. In the meantime, we are beginning to understand that increased fossil fuel consumption, unfortunately, leads to consequences we may not be able to undo. A changing climate and thus, deteriorating public health is one of the consequences.

Air pollution: As previously explained, fossil fuels result from dead organisms (animals & plants) buried deep inside the earth over millions of years. Under the right conditions of extreme heat, pressure and oxygen-free environment, these buried organic materials are turned into a solid rock where fossil fuels (coal, crude oil and natural gas) are extracted from.

Chinese air pollution Delima (Yantai, China)

Coal helped propel the Chinese economy to the 21st century. It fuels up to 78% of electricity generation and 60% of all energy use today. The consequences are that most of Chinese large cities are struggling with air pollution. Multiple studies have concluded that over 1 million Chinese die annually due to air pollution.

These decomposed organisms contain high levels of carbon and hydrogen molecules, which are the source of fuel/energy. Burning these fuels creates a chemical reaction between the carbon, hydrogen molecules and the oxygen in the air in a process called combustion. Combustion breaks the bonds between those molecules, which results in the energy we desperately need. Unfortunately, we get more than the energy we desperately need. Some of the carbon molecules mix with oxygen in the atmosphere to produce deadly byproducts of carbon dioxide (CO^2) and carbon monoxide (CO). These byproducts are the greenhouse gases that are causing much of the climate degradation. Since natural processes can only absorb about half of the amount of CO^2 that is emitted, there is a growth of net increase of atmospheric CO^2 annually. In the United States, the burning of fossil fuels, particularly from the power utilities and transportation industry, accounts for about three-quarters of carbon emissions.

In addition to the carbon based pollutants, burning fossil fuels releases other harmful agents. For example, coal-fired power plants singlehandedly generate about 40% of the dangerous mercury emissions and about 60% of the sulfur dioxide emissions in the U.S. Meanwhile, fossil fuels powered vehicles are the main contributors of poisonous carbon monoxide and nitrogen oxide, which are the causes of smog. Some of the most harmful agents emitted by coal combustion include:

- *Sulfur dioxide ($SO^{2)}$*: Reacts with water and oxygen to form acid rain.
- *Nitrogen oxides (NOx)*: Contributes to smog which leads to respiratory illnesses and cardiovascular effects. It is more dangerous to the elderly, young children and people with asthma.
- *Mercury and other heavy metals*: When deposited in soil and water, they lead to neurological and developmental damage in humans and other animals.
- *Fly ash and bottom ash*: Residues created by coal burning plants. They lead to lung cancer.

Coal-fired powered plants are recognized from a distance by their smoke/pollution towers

For more than a century, burning fossil fuels has generated most of the energy required to propel our cars, power our businesses and keep the lights on in our homes. Even today, oil, coal and natural gas provide about 80 percent of our energy needs. However, increasing fossil fuel consumption is contributing to an enormous toll on humanity and the environment. More importantly, the pollution released into the atmosphere has a much broader global reach and impact on public health.

Land degradation: In addition to air pollution, the process of extracting fossil fuels has a heavy toll on land and conservation. The required infrastructure to extract, process and deliver fossil fuels to end-users has a high level of destruction on conservation and surrounding communities. Mining, drilling and unearthing large acreage of land takes an enormous toll on the landscapes and the wildlife that depend on them.

For example, strip mining or surface mining for coal, involves blasting away the entire landscape including mountaintops, to expose below surface coal. Such is the process involved at the largest U.S. surface mine in Wyoming's Powder River Basin, which was the source of about 60% of the coal mined in the United States in 2019. Despite the best reclamation attempts, the land is never again suitable for previous native wildlife habitats.

Coal is also mined from underground using heavy machinery to cut coal from deep underground deposits. The process requires substantial financial investment and human health sacrifices to extract. It requires workers to be in an underground environment for an extended amount of time exposed to dangerous air particles. As a result, lung disease is widespread among coal workers.

In the case of crude oil, some of it is found in deep underground reservoirs where it is accessed by drilling, while some are located near the earth's surface in tar sands where it is accessed by strip mining. Thanks to modern technology, some of the crudes are also drilled from oceans and seas. Drilling in oceans, unfortunately, leads to shore contamination during oil spills, as witnessed in the BP oil disaster of the Gulf of Mexico in 2010.

While certain crude oil can be extracted using pumpjacks, offshore rigs or surface mining, others are too difficult or expensive to extract. Thus, they require a different extracting technique, such as hydraulic fracturing, also known as fracking.

Fracking is a method of extracting oil and natural gas from deep underground. The process involves the high pressure injection of fracking fluid containing water and other chemicals into a wellbore to create cracks in the deep-rock formations. When the hydraulic pressure is removed from the well, small grains of hydraulic fracturing proppants, such as sand, hold the fractures open and allows natural gas, petroleum and brine to flow upwards more freely. A few days later, the well is ready to produce oil or natural gas for years.

Unfortunately, fracking is also highly controversial. The environmental and public health impacts outweigh its benefits. Groundwater and surface water contamination are significant concerns. Residents, who live near fracking projects, have suffered health issues, such as pregnancy and congenital disabilities, migraine headaches, chronic rhinosinusitis, severe fatigue and asthma exacerbations. Fracking is also believed to initiate the triggering of earthquakes. For these reasons, fracking is under international scrutiny. In fact, it is restricted in some countries while completely banned in others.

Surface and Underground Coal Mining

Top: Comparison of blasted area compared to a natural landscape in the background.
Middle: The landscape will never recover. In fact, all the extracting chemicals and waste products will accumulate to create an acid or toxic lake.
Bottom: Underground mining are known for their safety and health risks

Methods of extracting oil

Pumpjack: A device used to extract crude oil from an oil well. An electric or gas-driven engine rotates counterweight to an arm that moves a pivoting beam up and down. Once at the surface, the crude oil is separated from any water and natural gas in the mix. The crude oil is then finally pumped into holding tanks before being transported to refineries. Pumpjacks are capable of pumping up to 10 gallons per stroke or up to 5 barrels per minute.

Oil platform: An oil platform, offshore platform or offshore drilling rig is a large structure with facilities for well drilling. The platform is used to explore, extract, store and process petroleum and natural gas that lies in rock formations beneath the seabed. Many oil platforms also contain facilities to accommodate their workforce. Most commonly, oil platforms engage in activities on the continental shelf, though they can also be used in lakes, inshore waters and inland seas. Depending on the circumstances, the platform may be fixed to the ocean floor, consist of an artificial island or float.

Hydraulic Fracturing (Fracking): Fracking is the process of extraction of natural gas or oil from rock formations deep underground. It requires drilling down into the earth and injecting the newly drilled well with high-pressure water, sand and chemicals mixture to release the gas trapped inside the rock formation and allow it to flow to the top of the well. The process can be carried out vertically or, more commonly, by drilling horizontally to the rock layer, which can create new pathways to release gas or to extend existing channels. According to the U.S. Environmental Protection Agency, fracture treatments in coalbed methane wells use from 50,000 to 350,000 gallons of water per well, while deeper horizontal shale wells can use anywhere from 2 to 10 million gallons of water to fracture a single well.

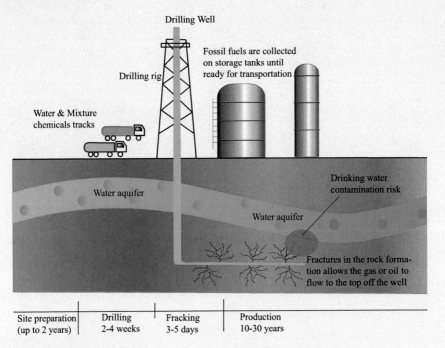

Simplified fracking process
It requires over 1400 truck trips to transport 2 to 5 million gallons of water

Oil/tar sands: are large deposits of bitumen or extremely heavy crude oil, consisting of a mixture of crude bitumen, silica sand, clay minerals and water. Extracting this crude oil involves large-scale excavation of the land with huge hydraulic power shovels and heavy hauler trucks. Below is the Athabasca deposit located in northeastern Alberta, Canada. This reservoir of crude bitumen is the largest known deposit in the world.

Water pollution: Coal, oil and gas development pose an undeniable threat to surface water and groundwater. The extraction process generates enormous volumes of wastewater, which can be laden with heavy metals, radioactive materials and other pollutants. For example, acidic chemicals runoff into streams, rivers and lakes during coal mining process. Oil is susceptible to spills during transportation, polluting drinking water sources and jeopardizing entire freshwater or ocean ecosystems. Since the start of the Industrial Revolution, the ocean has become 30 percent more acidic. The increased acidity of the oceans impacts entire food chains and the coastal communities that rely on them. The oceans compromise two-thirds of the earth's surface and absorb as much as a quarter of all carbon that are emitted to the atmosphere. Changing their chemical characteristics has enormous consequences that can be felt globally.

Following extraction, transporting the fuel to processing plants, refineries or end-users requires massive pipelines and road projects that impact many communities along the way. Additional transporting methods by train, trucks or supertankers expose the fuel to accidents and oil spills. For example, transporting Bakken oil from North Dakota fields to Irving Oil Refinery in New Brunswick, Canada, saw an estimated increase from 5000 train cars in 2006 to over 300,000 cars in 2012.

Mining Toxins

The process of extracting fossil fuels is an extremely dangerous process for both the people involved and the natural habitat of the area. Industrial explosives used to blast-open mining sites completely change the landscape beyond recovery. Chemicals used in the mining and drilling processes contaminate the land, water and air, causing health problems for workers and people living near mines. Some of the toxic chemicals used in mining and drilling fossil fuels include:

Sulfuric acid: is a byproduct of many kinds of mining operations. It mixes with water and heavy metals to form acid mine drainage or acid lakes. In addition, the sulfur oxides emitted from burning coal, react with moisture in the air to produce acid rain. Acid rain or any contact with sulfuric acid causes skin burns, blindness and death.

Mining operations produce acid byproducts. When mixed with water, they lead to acid lakes.

Hydrochloric acid (HCl): is used in fracking as part of a mixture of water, proppants and chemicals. The mixture is pumped into the rock or coal formation to dissolve some of the rock materials, to clean out pores and enable gas and fluid to flow more readily into the well. Like most acids, HCl is corrosive and causes damage to the skin upon contact. In addition, inhaling HCL causes eye, nose and respiratory tract irritation and inflammation, to name a few.

Ammonium nitrate and fuel oil (ANFO): ANFO is a mixture of 94% ammonium nitrate (AN) and 6% fuel oil (FO). It is a highly explosive compound widely used in blasting open pit coal mining.

Cyanide: A cyanide solution is used to extract gold from ore. Cyanide leaching allows profitable mining of much lower ore grades. Lower grade ore operation requires the extraction and processing of much more ore, which requires larger mines and larger open pits. Cyanide solutions easily flow into waterways and contaminate underground water supplies and agricultural lands, leading to substantial environmental and public health risks. Cyanide is a highly toxic and deadly element that can easily lead to death if accidentally swallowed.

Loss of Precious Spaces

Another consequence of increased fossil fuels consumption is the desperate search for additional reserves. The majority of the already known fossil fuels reserves are located in politically unstable regions of the world, such as the Middle East, North Africa and South America. Thus, nations are going to extreme measures to acquire additional or new supply of fossil fuels, often leading to conflict between industries, environmental groups, governments and local communities. Some of the most extreme measures include drilling in protected and precious areas.

Theodore Roosevelt National Park is a magnificent work of nature located in the U.S. State of North Dakota. It encompasses over 100 square miles of scenic drives, foot and horse trails, as well as hiking and camping. According to U.S. National Park Services, it is also home to many wildlife, including bison, bighorn sheep, white-tailed deer, mustang horses, elk and over 186 species of birds.

A few years after the park was established in 1947, oil was discovered in the Bakken shale formation, covering Montana, North Dakota (bordering the park) and the Canadian territories of Saskatchewan and Manitoba. The Bakken oil field is the largest continuous oil field in the world but was economically unprofitable to drill before 2006. The introduction of hydraulic fracturing/fracking technology changed the fate of this majestic territory. Estimates from a 2013 United States Geological Service (USGS) survey showed that the Bakken oil field could produce up to 11.4 billion barrels of recoverable oil. By the end of 2014, it was nearing one million barrels per day, contributing to 10% of U.S. total oil production, second only to Texas.

Despite the benefits of additional oil sources to the nation and the economic benefits of new jobs and tax revues to the local economy, the disastrous impact on the park is unmistakable. The development has occurred so quickly that the long-term social and environmental costs were unforeseen. Although the park is protected from oil drilling, the land just outside its boundaries is not.

According to the National Parks Conservation Association (NPCA), a rapidly increased production in the region has led to air pollution, water pollution, oil spills, illegally dumped fracking waste, habitat fragmentation, heavy traffic, impaired views and a host of social impacts. As many describe it, a state once known for its sparse population and ranching is now a giant oil industry playground. Fracking has allowed drilling companies to reach areas they were not able to with traditional vertical drilling. Fracking also produces large volumes of natural gas out of the same wells as the oil. With no infrastructure to process or carry away that natural gas, oil companies had chosen to either leave it mixed in with the oil and load it onto trains or burn/flare the potent greenhouse gas into the atmosphere. For a very long time, North Dakota was flaring about a quarter of the gas produced via fracking.

Following the coronavirus epidemic and oil price drop at the beginning of 2020, oil production from the Bakken oil fields of North Dakota fell by almost 50% to 827,000 barrels per day, continuing a production decline that started years ago. Operating oil wells have dropped to below 800 from their March 2020 peak of over 13,000. Estimate for the break-even oil price for drilling Bakken wells ranges from $38 to $60 per barrel. At a current fluctuation oil price of $50-$60 a barrel, North Dakota may have already seen its best oil production years. Now, the reconservation process starts. The large volumes of radioactive waste produced from fracking and the abandoned oil wells have to be cleaned up. Oil companies and Wall Street financiers have made their quick profits and are already looking to other unspoiled areas, such as the Arctic wildlife refuge in Alaska. In the meantime, North Dakota recently decided to use $66 million in federal funds designated for coronavirus relief to begin the cleanup efforts.

Dakota Access Pipeline or Bakken Pipeline: Drilling in an isolated and protected area is half the battle. Transporting the oil to the population centers is another challenge for all parties involved, including industry, government and local communities. Motivated by high oil prices, the oil industry was committed to getting the oil out of the Bakken ground and to customers as fast as possible. Since there are no oil refineries or transport hubs near the Bakken oil fields in North Dakota, moving oil by rail became the transportation of choice, leading to what is known as the "Bakken oil-by-rail" boom, which led to many oil spill accidents. For example, on July 6, 2013, one of the unit cars on a train carrying 77 tank cars, full of highly volatile Bakken oil, derailed and exploded in Lac-Mégantic, Quebec, on its way from North Dakota to the Irving Oil Refinery in New Brunswick. It destroyed nearly half of the downtown center and spilled a large amount of oil as it burned for days. Most importantly, 47 people lost their lives.

Alternatively, the U.S. Army Corps Engineers (USACE) was engineering an underground pipeline to transport the Bakken oil to refineries near Patoka, Illinois. Known as the Dakota Access Pipeline, the pipe is 1,172 miles long. It is designed to carry up to 570,000 barrels per day of crude oil beneath the Missouri and Mississippi Rivers and under part of Lake Oahe near the Standing Rock Indian Reservation. The Missouri River is also the primary drinking water source for the Standing Rock Sioux, a tribe of around 10,000 with a reservation in the central part of North and South Dakota. Siting a report from the Pipeline and Hazardous Materials Safety Administration (PHMSA) where more than 3,300 incidents of leaks and ruptures at oil and gas pipelines have been recorded since 2010, it is understood why local communities were concerned. Additionally, the construction path threatens many ancient burial grounds and cultural sites held sacred by the Sioux Nations and other neighboring Native American tribes.

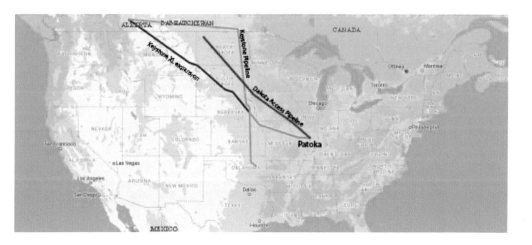

On January 24, 2017, President Donald Trump signed an executive order that reversed the Obama legislation and advanced the pipeline's construction. The executive order expedited the environmental review, which Trump described as an "incredibly cumbersome, long, horrible permitting process." On February 7, 2017, the Trump administration authorized the Army Corps of Engineers to proceed, ending the environmental impact assessment and the associated public comment period. The pipeline was completed by April and its first oil was delivered on May 14, 2017. Three years later, in July 2020, a District Court judge issued a ruling for the pipeline to be shut down and emptied of oil pending a new environmental review, only to be overturned by a U.S. appeals court a month later. At the timing of writing this book, the newly elected Biden administration has managed to shut down the pipe pending further investigation. If there is a lesson to be learned from this chaotic process, the continuous expansion in search of fossil fuels is environmentally, socially and politically toxic.

Arctic National Wildlife Refuge: The Arctic National Wildlife Refuge (ANWR) is the largest national wildlife refuge in the United States. It is located in northeastern Alaska and encompasses over 19 million acres. The region first became a federally protected area in 1960. This vast refuge of coastal lands, boreal forests and alpine tundra supports many species of plants and animals. Polar bears, grizzly bears, black bears, moose, caribou, wolves, eagles, lynx, wolverine, marten, beaver and migratory birds rely on this refuge. It is one of the finest, still intact, landscapes left on earth. Threatening this wildlife refuge is the estimated 7 to 11 billion barrels of oil stashed

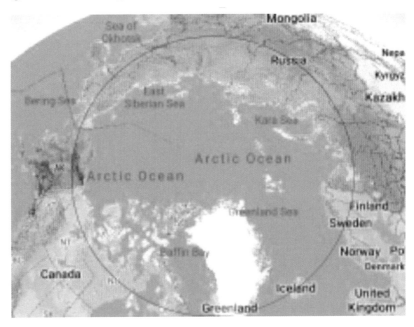

beneath it. A mere amount compared to Saudi Arabia's oil reserves worth over 87 billion barrels plus much more untapped proven reserves.

In December 2017, Congress passed the Trump administration's Tax Cuts and Jobs bill, which included a backdoor provision to approve lease sales for drilling in the refuge. This move allowed for oil and gas rights to be auctioned off in the heart of one of the world's most iconic wild places. Since the beginning of the Trump administration, over a million acres have been leased. According to the U.S. Bureau of Land Management, the millions of acres the Trump administration has offered for oil and gas drilling included sensitive wildlife habitat around the Teshekpuk Lake Special Area, which is one of the largest and most ecologically significant wetlands in the world.

Additionally, the administration put efforts to open up the entire 1.5 million acres of coastal plain for gas and oil exploration in September 2019. Although this area was added to the refugee protected status in 1980 by Congress with a mandate to study the petroleum reserves of the area, it is home to hundreds of species of birds, polar bears and the Porcupine caribou herd, which is a vital resource for the native Gwich'in people. The decision is scientifically and economically inexcusable. Fortunately, regulators have yet to finalize the environmental review process and the scheduled lease sale for 2019 has been postponed.

There is no doubt that the harsh weather conditions of that area would cause an oil spill. The U.S. Department of the Interior has concluded a 75 percent chance of a major oil spill if oil production resumes. Unfortunately, the studies have not stopped major companies like Shell and Exxon from aggressively pursuing a new "oil rush" in the Arctic Ocean. In fact, in parts of the Russian arctic, oil exploration has already begun. Russian oil giant Gazprom started producing oil from the region as early as 2013.

The Trump administration initially projected that leasing would generate $1.8 billion in revenue over a decade, but it has subsequently cut that estimate in half. Some local politicians hailed the decision as an economic boost for their state. They boasted that the oil boom will lead to new jobs and support economic growth and prosperity. These are short sited and short-term gain decisions at a time when oil prices are plummeting and oil consumption is losing ground to other sources rapidly. Major U.S. financiers like Goldman Sachs and JPMorgan Chase have already publicly announced their discontent and promised not to fund oil and gas projects in the Arctic refuge.

Regarding offshore drilling, President Obama issued an executive order to ban drilling in the Arctic offshore waters permanently. As expected, the Trump administration issued an executive order overturning the decision. Trump's decision was later overruled by Alaska federal Judge Sharon Gleason for being unlawful. A small victory for many in a long, messy war. There is no convincing argument for drilling in one of the most precious habitats on earth at a time when fossil fuels demand is on a decline. The largest economy in the world should be able to find $1.8 billion over a decade in some other ways without destroying one of the most precious and protected spaces on earth.

On the other side of the globe, the Russians and Europeans are a threat to the Arctic as they also plan to benefit from its resource. The Arctic is one of the world's fastest-warming regions, according to many scientists. This surge in climate change is destabilizing permafrost that has remained frozen year after year. As the permafrost destabilizes, so will the infrastructure (buildings, oil and gas pipelines, roads, railways and military bases) built on top of it. The incident experienced by metals giant Norilsk Nickel in May 2020 at its power station about 200 miles north of the Arctic Circle, is a good example. Melting permafrost shifted the foundation and ruptured a fuel reservoir, sending 21,000 tons of diesel into a fragile ecosystem of rivers and wetlands.

Protected areas serve undoubting significant importance to all generations. They are a few reminders of our planet's beauty. Unfortunately, their values are ignored for a short-term economic gain as they are being over-exploited for natural resources, especially oil. Resource exploitation of these areas may have been necessary at some point in the past. It is not necessary today. There are cleaner, cheaper, renewable and more efficient options. To those who believe these protected spaces can be developed responsibly, below are reminders of the dangers and the cost of fossil fuels exploitation. Regardless of how careful we are, equipment gets old and gets damaged, people make mistakes and sometimes the weather takes a turn for the worst.

Arctic Northern Lights. A natural wonder

Worst oil spill disasters

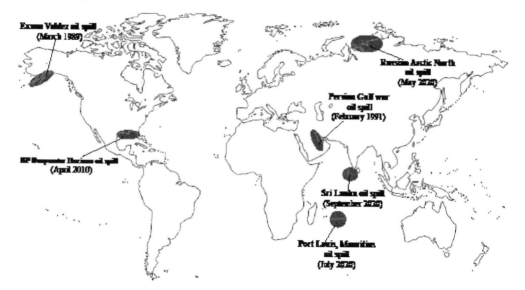

Exxon Valdez oil spill: On March 24, 1989, the oil tanker Exxon Valdez, bound for Long Beach, California, struck a reef and spilled nearly 11 million US gallons of crude oil in Prince William Sound, Alaska. The Valdez spill is the second-largest in U.S. waters and is considered the most environmentally damaging oil spill worldwide. The oil affected 1,300 miles (2,092 km) of coastline. Exxon spent US$2 billion on cleanup efforts but only recovered less than 7 percent of the oil spilled.

(Top) Exxon Valdez oil spill.

BP Deepwater Horizon oil spill: The largest accidental oil spill in history occurred in the Gulf of Mexico on April 20, 2010. A surge of natural gas blasted through a cement well cap traveling up the rig's riser to the platform. The accident killed 11 workers and injured 17 more. Environmentally, over 130 million gallons of oil were released (according to the findings of the U.S. District Court) and about 1,300 miles

(2,092 km) of the U.S. Gulf Coast from Texas to Florida were coated with oil. The oil platform capsized and sank several months later on September 17. In the lawsuits that followed, BP was found to be the responsible party and ordered to pay $65 billion in compensation to people who relied on the gulf for their livelihoods.

The Persian Gulf War oil spill: The world's largest non-accidental oil spill took place on August 2, 1990, following Iraq's invasion of neighboring Kuwait. After a massive air and ground campaign by the U.S. and coalition forces in January and February 1991, Iraqi forces responded by igniting hundreds of Kuwaiti oil wells. In addition, Iraqi troops released an estimated 400 to 500 million gallons of oil from Kuwait's Sea Island terminal into the northern Persian Gulf.

At the time of writing this book, we witnessed two oil spill disasters in Mauritius and off the coast of Sri Lanka within a month of each other.

Port Louis, Mauritius: The bulk carrier MV Wakashio ran aground on a coral reef off the southeastern coast of Mauritius on July 25, 2020, spilling over 1,000 tons of oil and threatening a protected marine park boasting mangrove forests and endangered species. Three weeks later, on August 15, a crack in a cargo hold at the vessel's stern forced the ship to break into two. Mauritius declared an environmental emergency and salvage crews raced against the clock to pump the remaining 3,000 tons of oil off the stricken vessel.

Sri Lanka: The New Diamond supertanker was burning for over a week, 34 miles off the coast of Sri Lanka. The supertanker, carrying the equivalent of about 2 million barrels of oil, had been transporting its cargo from Kuwait to a port in India when the fire broke out on Sept 3, 2020. Most of the fuel oil from the MT New Diamond was quickly contained, averting a major environmental and economic disaster for the 2 million Sri Lankans who depend directly on coastal fisheries and for the Sri Lankan tourism industry. More than 80% of hotels in Sri Lanka are built along the coast.

Russian Arctic North oil spill: A senior Russian official stated, "Diesel oil from a huge spill in Russia's Arctic north has polluted a large freshwater lake and there is a risk it could spread into the Arctic Ocean. Emergency teams are trying to contain the oil, which has now traveled about 20km (12 miles) north of Norilsk from a collapsed fuel tank."

According to officials, the oil started leaking on May 29th, 2020. Over 21,000 tonnes of oil had been released, contaminating the Ambarnaya river and surrounding subsoil. It is believed that the storage tank near Norilsk sank because of melting permafrost, which weakened its supports. The Arctic has had weeks of unusually warm weather, probably a symptom of global warming. The oil spill is considered the worst accident of its kind in modern times in Russia's Arctic region.

The power plant where it happened is run by a subsidiary of Norilsk Nickel, the world's leading nickel and palladium producer. Lake Pyasino serves as the basin for the Pyasina river, which flows to the Kara Sea, part of the Arctic Ocean. From October to June, that river is usually ice-bound.

Oil and toxic chemicals spill turned the river red

National Security and Global Conflict

All wars are resource wars: In addition to environmental concerns and loss of precious spaces, growing numbers of global conflicts are another consequence of increased fossil fuels consumption. Global population growth and rising consumerism are placing significant and potentially unsustainable pressures on the availability and sustainability of natural resources. Since fossil fuels are finite resources, their limited availability has been a source of increased global conflicts. According to the United Nations Environment Programme (UNEP) 2010 report, over the past 60 years, 40 percent of civil wars can be associated with natural resources. Additionally, there have been at least 18 violent conflicts fueled or financed by natural resources since 1990. The Secretary-General's Special Representative for the Democratic Republic of Congo (DRC), Martin Kobler, acknowledged in 2014 that the exploitation of natural resources had fueled the extensive conflict that has ravaged the country and taken millions of lives. The extraction of minerals such as Colton and Cassiterite for the electronic industry as well as gold, timber and oil has fueled conflicts both within the country and with neighboring countries.

Of course, resource conflicts go as far back as humanity itself. The earliest recorded resource conflicts date back to 4500 BC in the Middle East over freshwater supplies from the Euphrates and the Tigris rivers located in modern-day Iraq. Similar conflicts remain today in the region among neighboring countries over water distribution of the Jordan River. The U.N. states that over the last 50 years, freshwater withdrawals have tripled worldwide, with agriculture accounting for 70 percent of all water consumption. In comparison, industry and domestic consumption account for 20 percent and 10 percent, respectively. Unfortunately, no number of negotiations or conflicts have solved this region's water issue for thousands of years.

Water is also used in the form of hydroelectric power to generate electricity. Currently, it accounts for nearly 16% of global and 6% of U.S. electricity production. Diverting or changing water flow for hydropower deprives downstream population of a vital source for their livelihood. At the time of this book's writing, a major conflict is brewing between Ethiopia and Egypt over damming the Nile River. Previous international agreements allocated the majority of the Nile water resource to Egypt and Sudan and completely ignored the rights of the other nine countries that the Nile touches. As a result, those nine countries, especially the Nile source countries of Ethiopia (source of Blue Nile) and Uganda (source of White Nile), have demanded revision to the allocations for a larger share of the Nile benefits. Establishing new levels of distribution and reaching a shared agreement between all eleven countries has been challenging. Understandingly, Egypt strongly opposes any new revision as it considers the Nile waters a question of "life or death" and even threatening a military action. Adding to the pressure is a population growth projection in the region. According to the United Nations, Egypt's population is expected to grow from 83 million to nearly 130 million, while Ethiopia's is projected to grow from 83 million to 174 million by the year 2050. The combined population of North and South Sudan is projected to grow from 42 million to 76 million in the same time frame.

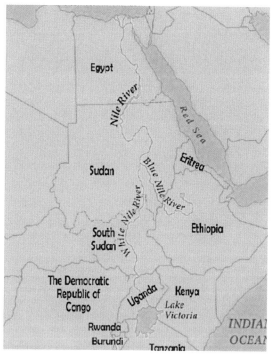

Nile River basin countries include all 11 countries where the river Nile and its branches touch.

Another heavily contested natural resource is the Forest, which covers about a third of the earth's landmass. The global forestry industry supplies timber and lumber for the building/construction industry, wood pulp for paper manufacturing and biomass feed for processing and power plants. The unceasing demand for wood and its products has led to extensive forest destruction globally and has had severe social and economic consequences for local communities. Driving the demand is the massive profits to be made, including an estimated annual $300 billion USD global market in the timber sector and an annual $60 billion USD market in the global paper manufacturing sector. The increased demand has also fueled many international illegal arms deals, bloody civil wars and regional instability. For example, timber revenue has funded conflict in northern Burma/Myanmar between the army and local militias trying to control the trade for decades. Similar conflicts are ongoing in Cambodia, Liberia and the DRC, to name a few.

And how can we forget the astronomical greed, loot and massacres portrayed by European powers over the last few hundred years in many parts of the world, including

Asia, Africa and the Americas to gain access to resources? Rich and fertile agricultural lands of Central and South America supplied European markets with tobacco, sugar and coffee. Vast lands of Asia were a source of spices, tea and silk, to name a few. Arguably, the richest continent on Earth, Africa, was a magnet for European conquest of the continent, including its people for labor. European powers went to war on all corners of the globe to gain and protect resources. Great Britain fought several wars with China to maintain and flourish its Opium market. The French fought brutally to keep Haitian plantations. King Leopold of the Belgians committed an indescribable horror in the DRC to keep wealth flowing to Belgium.

Whatever the commodity of the day, nations went to war to secure access to it. Today, since oil is the commodity of the day, the most prosperous oil region in the world, the Middle East, is and has been the hotbed of resource conflicts. Commercial oil was first discovered in the region in the early 1900s. Anglo-Persian Oil Company (APOC), later British Petroleum (BP), discovered and then started producing oil in Iran by 1911. Following World War I, APOC found more oil in neighboring Iraq. In 1932 Standard Oil Company of California discovered oil in commercial quantities in Bahrain, followed by discoveries in Saudi Arabia in 1938. World War II delayed the development of whatever fields had been discovered in the 1930s until the 1950s. During the 1950s and 1960s, additional regional countries, including Kuwait, Qatar and Abu Dhabi, started producing and exporting oil.

A few decades later in 1951, BP's investment in Iran came under attack following the election of a new prime minister, Mohammad Mosaddegh. Mosaddegh introduced a range of social and political reforms, of which the most significant was the nationalization of the Iranian oil industry, which had been built and operated by the British since 1911 through APOC/BP. Bitter over the loss of their control of the Iranian oil industry and determined to keep cheap oil flowing to western economies, the British along with the help of the U.S. government, overthrow prime mister Mosaddegh in a bloody coup in 1953. Iran's new government soon reached an agreement with foreign oil companies to restore the flow of Iranian oil to world markets in substantial quantities, giving the United States and Great Britain the lion's share of the restored British holdings. In return, the U.S. massively funded the resulting Mohammad Reza Shah's government until 1979, when the regime was toppled by a revolution that created an Islamic Republic and nationalized Iran's oil reserves again. Relation between Iran and Western countries has been toxic since.

Thousands of miles to the west, the Arab-Israel conflict tested Western powers' relationship with the Middle Eastern nations. In retaliation for Israel's invasion of Arab lands and in retaliation for U.S. assistance to Israel's war efforts, the Organization of Petroleum Exporting Companies (OPEC), which were made up of primarily Arab nations, agreed to cut back their oil production and supply to the world in 1970. Widely known as the "1970 oil embargo", it brought many world economies to their knees as oil was weaponized for political gain. The oil embargo of 1970 and the 1979 Iranian revolution would lead to oil market fluctuations that crippled the U.S. and most western countries' economies. The events in the Middle East taught the U.S. and its allies a lesson they would not forget. They were convinced that Energy insecurity equals National insecurity. In the decades to follow, the U.S. relied on its military dominance to secure resources worldwide.

For example, in 2001, the United State's ambition to topple seven regimes, including Iran, Iraq, Syria, Lebanon, Libya, Sudan and Somalia, was confirmed by former U.S. General Wesley Clark years later. All seven countries have a considerable amount of oil wealth or serve as strategic oil shipping lanes. Currently, the U.S. has maintained direct or indirect military presence from the mountains of Afghanistan to the deserts of Saudi Arabia, Iraq and Iran and to the shores of Syria to secure natural resources. The region that has recorded some of the earliest resource conflicts over freshwater over 6000 years ago continue to suffer from another resource conflict over oil.

Similarly, the South American nations of Venezuela and Ecuador continue to struggle with internal and external pressures for their vast oil resources. Venezuela, after all, has the largest oil reserves in the world. African countries such as Sudan and the DRC, as well as territories like the "South China Sea" remain hotbeds of conflicts for oil and natural gas. Even one of the most precious and isolated areas of the globe, the Arctic, has become a conflict zone between the United States, Russia and European nations.

Looking back at history and current global events, it is hard not to conclude, "All wars are resource wars". Since the creation of the United Nations in 1945, there have been many interstate and intrastate treaties to minimize global conflicts. In addition to the treaties, technology, along with favorable government policies and public awareness, can play a crucial role and contribute to a peaceful resolution of some of the global conflicts. For example, prioritizing energy supplies based on domestically available renewable sources is economically beneficial, reduces dependency on foreign nations and contributes to energy stability.

In addition, energy self-reliance through technological advances has an increasing contribution to global conflict resolution. For example, water desalination brings fresh drinking water to communities, hence, reducing conflicts among neighboring nations. The state of Israel gets over 50% of its drinking water supply from this technology. Additional technological solutions include:

- Responsible use of fertilizers and improved water management to minimize required water resources for agricultural output.

- Advanced construction techniques to minimize forest destructions for timber/lumber and thus minimizing conflicts for market share.

- Utilizing crops, such as corn and soybean, to replace wood biomass and preserve forests in the process. We can all benefit from leaving the forests of the world intact.

- Advancing renewable energy, such as wind and solar, to meet the energy demands of a fast-growing population and to replace resources that have previously led to conflicts.

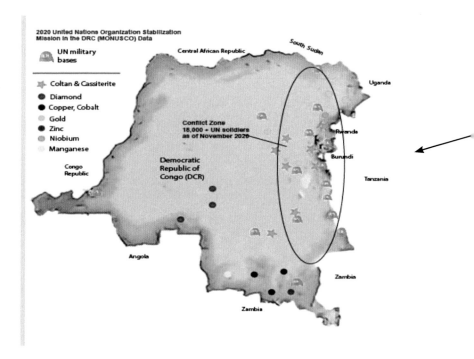

Resource conflict in the DRC

The DRC is the 2nd largest country in Africa and the 11th largest in the world. It comprises an area as large as all of western Europe. The DRC is also arguably the richest country in the world. It is blessed with all kinds of natural resources such as copper, gold, diamonds, cobalt, uranium, coltan, rubber, oil and thick forests. Limitless water, from the world's second-largest river, the Congo, great climate and rich soil make the DRC fertile as well.

Unfortunately, the DRC is also home to the world's bloodiest conflict since World War II. Over five million people have died and millions more have been driven to the brink of starvation and disease. Since gaining independence in 1960, there have been countless armed conflicts involving over a hundred internal and external armed groups in the country.

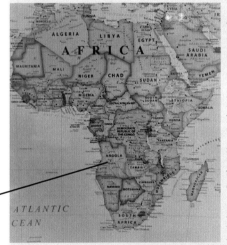

All the country's commercial ports and political power seats are on the western side of the country along the Congo river and the Atlantic Ocean. Yet, all of its armed conflicts and military unrests have been and still are, on the less populated eastern side. It is also no secret that the majority of the country's resources are located in the eastern side of the country. After all, wherever there are resources, armed conflicts are almost guaranteed.

Ever since the Europeans arrival on its shores, the DRC has seen nothing but slavery, death, brutality and impoverishment. When Portuguese traders arrived from Europe in the 1480s, they realized they had stumbled upon a land of vast natural wealth, including natural and human resources. Soon, the Congo was home to a supply of slaves and about four million people were forcibly shipped to the Americas on English ships. Late in the 19th Century, the interior of the Congo was opened up by the British-born explorer Henry Morton Stanley. Soon after, King of the Belgians, Leopold II, would hack a vast part of the country as his private empire. Under his control, millions of Congolese would perish. Today the country remains one of the poorest in the world and its citizens have not experienced peace and stability for centuries. Unfortunately, the DRC's resource blessings have turned into curses.

The Price of Change

There is too much scientific evidence to turn a blind eye to climate change. Therefore to act accordingly and responsibly is the only option. Acting responsibly means cutting back on fossil fuels consumption and gradually shifting towards renewable energy sources. But at what cost? Many fear that harsh policies and regulations on fossil fuels consumption mean losing opportunities of maintaining a stable livelihood. The fossil fuel industry has been responsible for significant economic success worldwide. As mentioned above, fossil fuels have helped launch the industrial revolution and propel our lives into the modern world. Investments in the fossil fuels industry have created economic opportunities and millions of jobs for many for the past 150 years. Thus, any attempt to suppress the industry and shift towards renewable energy sources must first address the below two questions:

- Can the renewable energy industry create as many jobs as the fossil fuels industry?
- Can the renewable energy industry compete with the fossil fuels industry on pricing?

Job Creation: Below is data from the U.S. Department of labor regarding job creation by the fossil fuels industry. In 2019, the industry was directly responsible for a little over 800,000 jobs. Indirectly, the fossil fuels industry contributed nearly 1.5 million jobs to the U.S. economy. Despite the 1.5 million plus jobs it creates, the fossil fuels job market is a very small portion of the overall U.S. job market. For example, the Department of Labor states that non-agricultural job market for 2019 was approximately 151 million. As shown in the table below, the most prominent sectors are government, healthcare, retail, manufacturing and services. The top 10 industries, which do not include the energy sector, make up over 85% of all jobs in the U.S. economy. The energy sector is responsible for a little over 6 million jobs or about 4% of all jobs in the market.

Most importantly, the renewable energy industry, even in its infancy, provides about half of the 6 million jobs in the energy sector. Thus, loss of jobs is not a valid argument against shifting towards the renewable energy economy. The following charts and graphs illustrate the contribution of the renewable energy industry in the overall U.S. job market.

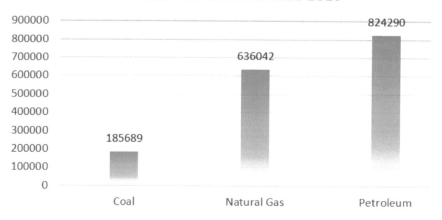

U.S. labor market trend

Industry	Employment (thousands of jobs)		
	2009	2019	2029
Total U.S. Employement	143,036	162,796	168,834
Nonagriculture wage and salary	132,029	151,710	158,116
State and local government	19,723	19,759	20,080
Health care and social assistance	16,540	20,413	23,492
Retail trade	14,528	15,644	15,276
Manufacturing	11,848	12,840	12,395
Accommodation and food services	11,162	14,143	15,024
Food services and drinking places	9,399	12,065	12,942
Local government educational services - compensation	8,079	8,010	8,033
Professional, scientific, and technical services	7,553	9,543	10,577
Administrative and support and waste management and remediation services	7,208	9,343	9,696
All other retail	7,094	7,489	7,278
Administrative and support services	6,857	8,888	9,210
Other services	6,150	6,714	6,995
Construction	6,017	7,492	7,792
Finance and insurance	5,844	6,425	6,491
Ambulatory health care services	5,793	7,697	9,124
Wholesale trade	5,521	5,903	5,801
Hospitals	4,667	5,199	5,454
Transportation and warehousing	4,225	5,618	5,944

A closer look at the energy labor market shines a light on the promise of the renewable energy industry. As mentioned previously, the energy sector contributes 6 million-plus jobs to the U.S. labor market. Below is a breakdown of these numbers.

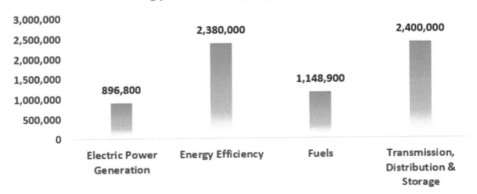

Electric power generation: 896,800 jobs

A more detailed look at the energy job market shows how the utility sector (electric power generation) has made a great stride in shifting from fossil fuels feed to renewable energy feed. The electric power generation sector employed about 896,800 Americans in 2019, with renewable energy contributing 68% of them (below chart). Solar and wind were the top two job-creating sectors of the renewable energy industry.

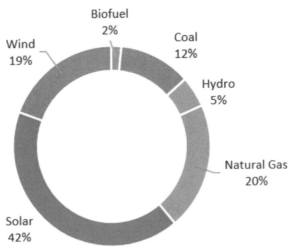

The Price of Change

Energy efficiency: 2,380,000 jobs

The energy efficiency sector is also another promising sector where a large percentage of the 2.3 million jobs are in the renewable energy sector. The Energy Star program alone creates jobs in energy-saving appliances and LED lights manufacturing, which will be discussed in detail in later chapters. Construction jobs involve jobs in solar panels and wind turbines manufacturing and installations.

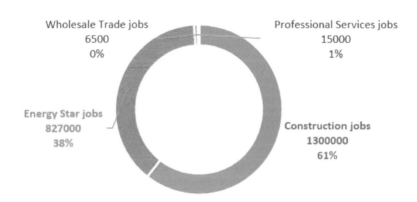

Fuels: 1,148,900 jobs.

The fuel/transportation sector has made great strides in converting from fossil fuels to renewables. Biofuel and biodiesel have become just as common, while Electric Vehicles (EVs) have been on the rise eliminating the need for fossil fuels altogether. This sector is the topic of other publications and will not be addressed in this book.

Transmission, Distribution & Storage: 2,400,000 jobs:

It is becoming more apparent that reliability in solar and wind energy will heavily depend on battery storage technology. As a result, this sector has also gained recognition as battery technology is finally coming to light for utilities and EVs.

Cost of Electricity: Many parts of the developed world are accustomed to cheap electricity prices. For example, in the U.S, the cost of electricity for residential, commercial and industrial consumptions are about 13 cents/11 cents/and 7 cents. These low prices are mainly driven by fossil fuels (coal & natural gas) which made up 62% of the electricity production and nuclear power, which was responsible for another 20% of all U.S. electricity generated in 2019. For the renewable energy industry to be taken seriously, the industry needs to provide electricity at a competitive cost without any government subsidies. After all, money is the ultimate decider.

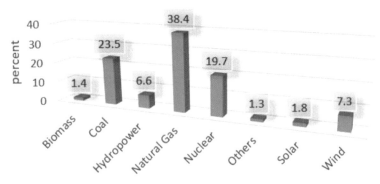

Non-renewable energy is responsible for 81% of U.S. electricity generation: 2019

Multiple researches have shown that not only can renewable energy compete with traditional fossil fuels on pricing, but it is actually cheaper. The below chart shows trends in the cost of electricity generation with renewable energy feed. For example, the price of generating electricity with solar power has gone down from 38 cents per kWh in 2010 to 3.9 cents in 2021, making it the cheapest source of electricity. Onshore wind power is the second cheapest at 4.3 cents per kWh. Both solar and wind power have become the lowest sources of electricity in a short ten years. Other sources, such as hydro, are not far behind at around 5 cents per kWh.

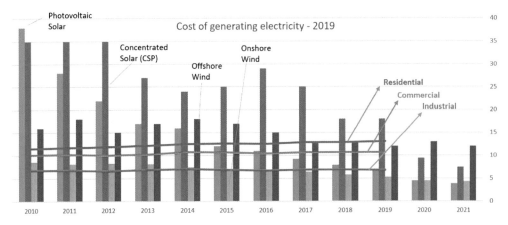

Cost of electricity generation. Declining trend for renewable sources

The Price of Change

There is finally an opportunity to support both sides of the climate change debate. An opportunity to shift towards a cleaner and renewable energy source while creating greater investment and economic opportunities for many. We have been witnessing a gradual shift from a traditional source of energy to renewable energy in the production of electricity for the past few decades. As shown in the graph below, the dominance of coal started to decline and be replaced by natural gas around 2017. Consumption of renewable sources, such as hydro, wind and solar, started to pick up around the same time.

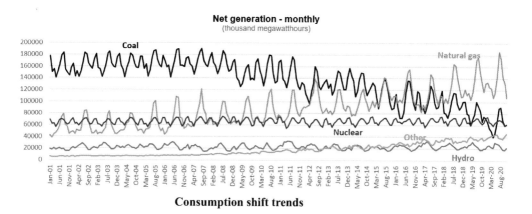

Consumption shift trends

Renewable energy has proven to be a cheaper source of energy. It has also proven to be a job creator, even in its infancy. As the industry develops, there is no doubt the job numbers would skyrocket, providing many with a cleaner and more lucrative source of livelihood. One reason for this success is the many different renewable energy sources available, including solar, wind, geothermal, biomass and hydro. These options allow many regions to capitalize on their locally available resources. For example, hot and arid climates can take advantage of the long hours of sunshine for solar power, while windy regions like the American midwest can benefit from wind power. Some areas like Iceland, benefit from geothermal energy while Brazil and India utilize their abundant source of biomass and hydropower.

In the chapters to follow, we will further discuss and elaborate on understanding the many different types of renewable energy and how they operate. We will start with one of the most promising sectors, which is the Energy efficiency sector. After all, energy we don't use is energy we don't have to generate.

Energy Efficiency

Out of the $1.8 Trillion dollars (USD) invested in total energy in 2018, $240 billion (13.3%) was invested in energy efficiency for buildings, transport and industry sectors, according to the International Energy Agency (IEA) 2019 report. The most significant portion of the investment, nearly $140 billion, was invested in the building sector. Transport energy efficiency, including heavy-duty and light-duty freights and electric vehicles (EV), received $60 billion, while industrial efficiency got an investment of $40 billion.

The massive investment has helped the energy efficiency sector create over 2.3 million jobs in the United States as of 2019. The sector includes ENERGY STAR appliances and lighting, pollution reduction and removal, greenhouse gas reduction, recycling, natural resources conservation and environmental compliance, and retail selling of energy-efficient products.

In the U.S., the jobs are dispersed across the country, with workers employed in all but seven counties, according to the Environmental and Energy Study Institute (EESI). More than 300,000 people are employed in energy efficiency in rural areas and 900,000 people work in energy efficiency in the country's 25 largest metro areas. Additionally, according to the U.S. Energy and Employment report, over 800,000 Americans are employed in manufacturing or installing ENERGY STAR certified appliances, including heating and cooling.

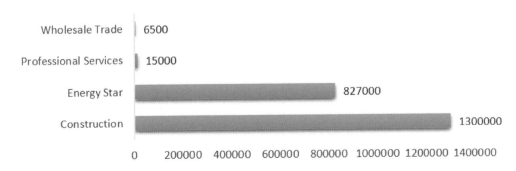

Building sector efficiency: Old techniques

The building sector is the largest in the energy efficiency industry. It incorporates old and new techniques to help save energy use for heating/cooling and powering homes and businesses. It starts with centuries-old techniques, such as passive solar energy, and incorporates technological advances in insulation techniques as well as intelligent appliances and accessories to achieve energy efficiency.

Ancient cultures have long used different methods to maximize the sun's energy for heating and cooling their habitats. For example, doors and windows were built facing the sun to allow heat in winter times, while different shading techniques were used to block the sun's heat during summer. In addition, building materials such as clay bricks and mud were used in construction for better temperature management. In some cases, homes were built underground or under earthy materials to take advantage of the minimal temperature fluctuation.

The Native Americans' cliff dwelling in the southwestern part of the U.S. is a perfect example of utilizing location, orientation and building material to help minimize the daytime and nighttime temperature fluctuation to create a comfortable living habitat in an arid and hot environment. This method of temperature control is called "Passive Solar Energy". It does not involve mechanical or electrical devices as in Active Solar Energy, which will be discussed in later chapters. Passive solar energy is used to either heat buildings (passive solar heating) or cool buildings (passive solar cooling) using two common techniques; building orientation and building materials.

Native American Cliff Dwelling at Mesa Verde, Colorado

Homes are built on the side of a cliff to limit direct sun exposure, thus, keeping the inside daytime temperature cooler. The earthy building materials also help block the sun's heat during the day by absorbing much of the heat. The absorbed heat is then gradually released into the interior at night when temperatures are much colder. This illustrates an excellent use of orientation and material know-how.

Modern structure in Texas, USA

Modern struture built with similar earthy building materials of stucco. The stucco absorbs the sun's heat during the day while releasing the heat during cooler night temperatures. The structure is built with many windows on all sides to allow heat in winter. During summer, however, the heat that penetrates through those windows is absorbed by earthy floor materials, such as tile floors. These tile floors serve the same purpose as the outside stucco in gradually releasing absorbed heat at night.

Passive solar heating is the concept of using the sun's energy to heat a living space. Passive solar heating system is utilized during colder fall or winter seasons to absorb as much of the sun's light and heat as possible during the day. This is accomplished by orientating openings such as windows, doors and sunroofs to face the sun. For example, in the northern hemisphere, such as North America and Europe, building doors and windows would face south (30 degrees of true south). In contrast, in the southern hemisphere like Africa and South America, doors and windows would face north to take advantage of the sun's energy. This orientation allows for maximum sun exposure during the cold winter days.

However, at night, when the sun's energy is absent, different techniques are required to create comfortable heated environment. Evenings and nights are normally colder in many parts of the world, even in hot arid climates. To combat the extreme temperature difference, many ancient cultures had selected their building materials very smartly. They used earthy building materials such as mud and clay to build their habitats. The concept is that these earthy materials have the capacity to absorb the sun's energy during the day and slowly release it back into the living space at a much slower rate, thus, heating the space throughout the night. This concept of selective building materials takes advantage of the materials' Thermal Mass.

Thermal mass is the material's capacity to absorb energy and release the absorbed energy gradually for use at a later time. Thus, building materials with excellent thermal mass absorb the sun's heat during the day and release it during cooler nights when the heat is much needed. This process helps reduce daytime and nighttime temperature fluctuation and provides more comfortable living space.

Building Material	Specific heat Capacity (J/kgK)	Density (kg/m3)	Effective heat Capacity (Wh/m2K)
Water	4200	1000	175
Cast concrete	1000	2000	83.3
Concrecte Block	840	2240	73.1
Brick	800	1750	42.4
Timber	1600	650	5.4
Ceramic Tiles	800	1900	4.2
Wet Plaster	1000	1330	3.7
Plasterboard	840	950	2.7
Stone	900	2000	
Earth Wall (Adobe)	837	1550	

All passive solar energy building materials are ranked according to their thermal mass. The material's thermal mass is categorized according to its heat capacity (how much it absorbs) and density (mass or volume). Heat capacity is the amount of heat energy required to change the temperature of an object. For example, it takes a lot of energy to change the temperate of concrete or clay compared to metal. Thus, concrete and clay have high heat capacity, which is desirable for passive solar design.

Additionally, the total amount of energy an object can store depends on its density or mass. Therefore, high-density materials like tile, concrete or brick absorb much more energy during the day and are more desirable for passive solar design. Table EE01 shows a list of materials and their thermal mass rankings.

High thermal mass building materials commonly used:

- Concrete, clay bricks and other forms of masonry: Concrete's thermal mass properties save 5-8% in annual energy costs compared to softwood lumber.
- Earth, mud and sod: In addition to using these materials for building, people sometimes use earth sheltering around their homes for the same effect. Basements are a great example.
- Logs or solid wood have significant thermal mass like concrete or earthy materials. In addition, solid wood has the added benefit of a much better insulation characteristic. For these reasons, log homes are common in colder climates. The insulation factor keeps the inside temperature insulated and warm during winter times, but the thermal aspect absorbs much of the daytime heat and helps keep the interior space cooler during summer.

Passive solar cooling: We have seen how passive solar heating concept uses building orientation and building materials to heat interior living spaces. However, cooling the same environment during hot summer days takes a different approach. In fact, in hot arid climates, such as the southwest region of U.S., where daytime and nighttime temperatures fluctuate extremely, moderating living space temperatures can be challenging. Fortunately, the same building materials used in passive solar heating can be used here as well. For example, during the summer heat, when the sun's energy is heating the buildings' walls, high thermal mass materials, such as stone and brick, will absorb the sun's energy rather than transfer it to inside the building, contributing to cooler temperate inside.

The energy that hits the outside walls is absorbed by the building materials while the energy that passes through the windows and doors is absorbed by the floor materials. The right materials help store the energy during the day to be released slowly throughout the evening & night to keep the home warm. This method is beneficial in areas where there is a large fluctuation between daytime and nighttime temperature.

Ancient Anasazi Indians in Arizona and New Mexico built their homes and villages on the side of cliffs to shade their homes from the sun and to take advantage of the natural cooling offered by the rocky cliff face. Other cultures like the Japanese and Vikings dug pit houses, also called sunken houses, to regulate their dwelling's temperature naturally. Below are samples from across the globe where earthy building materials were used similarly.

Ancient styled homes in two different continents. Both are built of mud or clay walls and a combination of wood and hay straw roofs. These materials help absorb the sun's energy during the day and help keep the interior temperature cool. The absorbed energy is slowly released at night to help keep the interior warm.
(Left) Eritrea, Africa (Right) Ireland, Europe

Although few of us are prepared to go back to ancient building techniques, we can use a more modern version of these techniques in our modern homes. Today, people use similar earthy materials such as stone, stucco and bricks, especially in warmer climates such as Southern Europe and South West United States. In other parts of the world, where earthy materials are not used, the chosen material is carefully prepared to provide the same "temperature regulating" benefits.

In humid climates, such as the Southeast United States, night temperatures are as humid as daytime, especially during summer. Thus, earthy building materials that help cool inside temperature by absorbing daytime heat, turn around and release the absorbed heat back into the interior space at night, even when the heat is not needed. This creates a much warmer internal temperature, possibly leading to overheating. Although this challenge can be solved with adequate ventilation at night to carry away stored energy, it is a clear demonstration of how "one size does not fit all". It is critical that we understand our local climate and have a great understanding of building materials to take advantage of nature's gifts and challenges. Coupled with the right building orientation, we can gain more control of our habitats' heating and cooling without burning our wallets.

Ancient material - Modern design

This Georgia, USA home is built of bricks on all sides. Brick has an excellent thermal mass and thus, absorbs the daytime heat and helps keep interior temperature cool. At night, however, the absorbed heat is released into the interior for warmer night temperature. During summer, when nighttime temperatures are still hot and humid, proper ventilation is used to carry away unnecessary energy.

Additional techniques for passive solar cooling systems are shading and natural ventilation, which help reduce unwanted daytime heat and keep cooler daytime and nighttime temperatures. Most importantly, we must do everything we can to block the sun's energy from entering our building in the first place. Several time-tested methods include:

- **Shades:** Trees that blossom during summer (thus blocking the sun) and shed their leaves during winter (thus allowing more sun to pass thru).
- **Overhangs:** Since the sun's path follows a high arc during summer,

roof overhangs are used to block the sun's rays and thus provide additional cooling.
- **Blinds:** Blinds give us the control of blocking or allowing the sun's energy.
- **Ventilation:** The easiest and most natural way to cool a space is ventilation: opening windows to circulate the air.

In addition, there are more advanced techniques and products of passive solar designs, such as window glazing/thermal mass ratio, that are becoming more common in modern windows and glass wall installations.

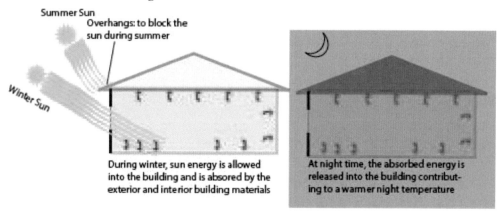

During winter, sun energy is allowed into the building and is absorbed by the exterior and interior building materials

At night time, the absorbed energy is released into the building contributing to a warmer night temperature

Trees play an excellent role in providing shades & blocking daytime heat. The tree in front of this home provides shade during summer. The leaves fall off in winter allowing more of the sun's energy to pass through the windows. The roof also has a slight overhang that helps block the summer sun.

Modern office building built with steel overhangs for summer shading. During winter, more light/heat passes through the windows since the fun follows a lower path arc.

Ancient Wisdom

(From Khartoum, Sudan to Colorado, USA)

I remember living in Khartoum, Sudan as a young teenager. Summer temperatures were scorching. Most families at the time didn't have the luxury of modern appliances, such as refrigerators. So families filled a clay pot full of water and buried the clay pot halfway into the ground. Oh! How cool the water was. I can still taste it.

Years later, growing up in Colorado, where temperatures varied from low teens during winter to dry and hot during summer, the basement was my favorite spot. Underground spaces are surrounded by earthy materials and thus have less temperature fluctuation between day and night. This makes basements cooler than the surface temperate in the summer and warmer than the surface temperature during winter. This makes the basement an ideal place year-round. It's incredible how we benefit from ancient techniques without even realizing it.

Building sector efficiency: New techniques

Insulation: In addition to the above-discussed methods, reducing the amount of air that leaks in and out of a home is a cost-effective way to cut heating and cooling costs, improve durability, increase comfort and create a healthier indoor environment. Insulation helps control airflows and can be applied either during new construction or on existing buildings. Fortunately, unlike the old insulation of hays and straws on rooftops, we have the luxury of insulation made from fiber, foam and other modern materials today.

Insulation levels are specified by R-Value, which is a measure of insulation's ability to resist heat flow. The higher the R-Value, the better the thermal performance of the insulation. For example, the attic is one of the most important and most accessible places to add insulation to improve comfort and energy efficiency for a home. The recommended level for most attics is R-38 or about 10 to 14 inches thick, depending on insulation type. Also, any walls that face the outside need to be adequately insulated. The proper insulation means less hot or cold air escapes the building, which means more money in our pockets. The map and table below illustrate the thickness of insulation required in certain parts of the home depending on the area of the country (categorized by climate zones).

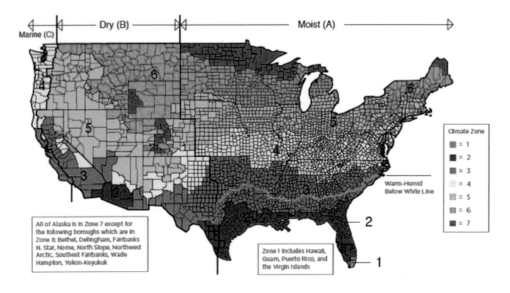

U.S. climate zones
Source: U.S. Department of Energy

Zone	Attic	2x4 wall	2x6 wall	Floor	Crawlspace
1	R30 to R49	R13 to R15	R19 to R21	R13	R13
2	R30 to R60	R13 to R15	R19 to R21	R13	R13 to R19
3	R30 to R60	R13 to R15	R19 to R21	R25	R19 to R25
4	R38 to R60	R13 to R15	R19 to R21	R25 to R30	R25 to R30
5	R49 to R60	R13 to R15	R19 to R21	R25 to R30	R25 to R30
6	R49 to R60	R13 to R15	R19 to R21	R25 to R30	R25 to R30
7	R49 to R60	R13 to R15	R19 to R21	R25 to R30	R25 to R30

Air leakage: Before applying insulation, especially in existing structures, detecting and measuring air leakage is economically smart. Air leakage occurs when outside air enters and interior heated or air-conditioned air leaves the home through cracks and openings. Detection can be accomplished visually or with measuring devices and techniques.

The recommended strategy is to reduce air leakage as much as possible and to provide controlled ventilation as needed. The United States Department of Energy has excellent recommendations to follow at www.energy.gov. It starts with a visual inspection of the outside of a home where there are joints or areas where two different building materials meet such as, exterior corners, water faucets, where siding and chimneys meet, and areas between the foundation and the rest of the building.

Inside the home, any cracks and gaps that could cause air leaks must be inspected. Those include doors and window frames, cabling outlets, gaps around pipes and wires, fireplaces, and attics. Today, we can benefit from technologically advanced windows and doors with higher insulation performance and glazing. However, cheaper solutions such as plastic sheets over windows are always better than nothing.

Once air leak areas are identified, caulking and weather-stripping are two simple and effective air-sealing techniques that offer quick returns on investment. Caulk is generally used for cracks and openings between stationary house components such as around door and window frames. On the other hand, weather-stripping is used to seal components that move, such as doors and operable windows.

Tips for Sealing Air Leaks (courtesy of the U.S. Department of Energy)

Following the detection of air leaks and the assessment of ventilation needs for indoor air quality, the sealing process can start as follows:

- Weather-strip doors and windows that leak air.
- Caulk and seal air leaks where plumbing, ducting or electrical wiring comes through walls, floors, ceilings and soffits over cabinets.
- Install foam gaskets behind the outlet and switch plates on walls.
- Inspect dirty spots in your insulation for air leaks and mold. Seal leaks with low-expansion spray foam made for this purpose and install house flashing if needed.
- Look for dirty spots on your ceiling paint and carpet, which may indicate air leaks at interior wall/ceiling joints and wall/floor joists, and caulk them.
- Cover single-pane windows with storm windows or replace them with

more efficient double-pane low-emissivity windows.
- Use foam sealant on larger gaps around windows, baseboards and other places where air may leak out.
- Cover your kitchen exhaust fan to stop air leaks when not in use.
- Check your dryer vent to be sure it is not blocked. This will save energy and may prevent a fire.
- Replace door bottoms and thresholds with ones that have pliable sealing gaskets.
- Keep the fireplace flue damper tightly closed when not in use.
- Seal air leaks around fireplace chimneys, furnaces and gas-fired water heater vents with fire-resistant materials such as sheet metal or sheetrock and furnace cement caulk.
- Fireplace flues are made from metal and over time, repeated heating and cooling could cause the metal to warp or break, creating a channel for air loss. To seal your flue when not in use, consider an inflatable chimney balloon. Inflatable chimney balloons fit beneath your fireplace flue when not in use. They are made from durable plastic and can be removed easily and reused hundreds of times. If you forget to remove the balloon before making a fire, the balloon will automatically deflate within seconds of coming into contact with heat.

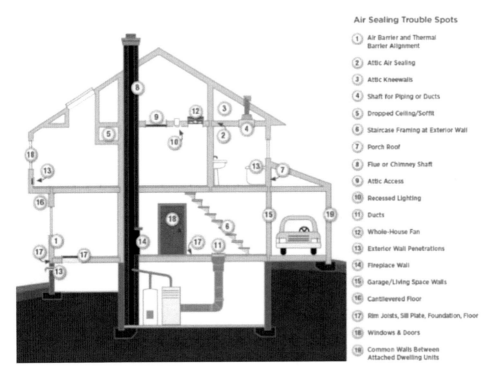

Energy star program

Modern appliances & accessories: Modern appliances and accessories play a significant role in optimizing a passive energy concept. In the United States, the U.S. Department of Energy (DOE) classifies these accessories according to their "Energy Star". Energy Star is a certification program for manufactures and suppliers to provide household products as energy-efficient as possible. These products help consumers save money on operating costs by reducing energy use without sacrificing performance. They also promote sustainability and reduce greenhouse emissions. According to the DOE, Energy Star and its partners helped American families and businesses save more than 4 trillion kilowatt-hours of electricity and achieve over 3.5 billion metric tons of greenhouse gas reductions since 1992. The greenhouse gas reduction is equivalent to the annual emissions of more than 750 million cars. In 2018 alone, Energy Star helped Americans save nearly 430 billion kilowatt-hours of electricity and save $35 billion in energy costs with associated emission reductions of 330 million metric tons of greenhouse gases. Energy Start products include appliances, lighting, electronic and data equipment, heating/cooling products, windows and doors. Americans purchased more than 300 million Energy Star certified products in 2018. Considering a typical American household spends $2,000 a year on energy bills, Energy Star products help save up to 30% or about $575 of the energy bill.

Energy star and real estate: DOE estimates that over 2 million Energy Star certified new homes and apartments are built to date, with nearly 100,000 built in 2019 alone. Today, constructing Energy Star certified homes has become a standard process for most of America's largest homebuilders. Energy Star certified homes and apartments are at least 10% more energy efficient. Existing buildings can also meet Energy Star certification by adding efficiency products such as insulation, windows and doors. Commercial and industrial buildings have a lot to gain from the Energy Star program as well.

On average, they waste 30% of the energy they consume. American businesses spent approximately $350 billion on energy costs in 2016 to operate commercial and industrial facilities. These businesses would save roughly $35 billion, with just a 10% improvement in energy efficiency.

U.S. Department of Energy Facts:

- The estimated annual market value of Energy Star products sales is more than $100 billion.
- On average, Energy Star certified buildings use 35% less energy than typical buildings nationwide.
- In 2018, the Energy Star program for commercial buildings helped businesses and organizations save 190 billion kilowatt-hours of electricity, avoid $12 billion in energy costs and achieve 140 million metric tons of greenhouse gas reductions.
- In 2018, the Energy Star program for industrial plants helped businesses save 36 billion kilowatt-hours of electricity, avoid $3 billion in energy costs, and achieve 40 million metric tons of greenhouse gas reductions.
- In 2018, the Energy Star Residential New Construction Programs helped homeowners save 3 billion kilowatt-hours of electricity, avoid

Times Square, New York, NY

$400 million in energy costs and achieve 4 million metric tons of greenhouse gas reductions.
- 2,800 builders, developers and manufactured housing plants are Energy Star partners, including all of the nation's twenty largest homebuilders. One out of every 12 single-family homes built in 2019 was Energy Star certified.
- Energy Star partners completed over 98,000 home improvement projects to increase energy efficiency and comfort in 2019, for a total of more than 873,000 to date.

LED lights: Americans spend over $300 billion on residential and commercial electricity bills annually, of which 15% is on lighting alone. As a result, many organizations, including the DOE, have heavily invested in lighting technology and policies to improve efficiency and reduce the cost of lighting, leading to the rapid adoption of LEDs.

LED stands for light-emitting diode. An electrical current passes through a microchip, illuminating the tiny light sources we call LEDs, and the result is a visible light. LED lighting products produce light up to 90% more efficiently than incandescent

Marketing and outdoor displays are taken to new heights as LED lights have reduced the cost of electricity by nearly 90%.

light bulbs and 80% more efficient than compact fluorescent lamps (CFL). With other lighting types, the light must be reflected in the desired direction and more than half of the light may never leave the fixture. LEDs use heat sinks to absorb the heat produced by the LED and dissipate it into the surrounding environment. This keeps LEDs from overheating and burning out. Good quality LED bulbs, especially Energy Star rated LED bulbs which are verified for quality and performance, can last more than 25 times longer than traditional light bulbs. For example, LED lights last up to 100,000 hours compared to 3,000 hours for incandescent lamps. In addition, LED lights are made from durable acrylic lenses instead of glass. This material makes LED lights much more resistant to breakage.

According to the November 2015 Department of Energy report, the U.S. had around half a million LED lights installed in 2009. By 2014, that number had increased to 78 million. Today, DOE estimates at least 500 million recessed downlights are installed in U.S. homes and more than 20 million are sold each year.

Advancements in technology and aggressive energy policies have helped reduce the cost of LED lighting by nearly 90% since 2009. By 2030, it's estimated that LEDs will account for 75% of all lighting sales. Switching entirely to LED lights over the next two decades could save the US $250 billion in energy costs, reduce electricity consumption for lighting by nearly 50% and avoid 1,800 million metric tons of carbon emissions. Because of their unique characteristics, including high efficiency, compact size, ease of maintenance, resistance to breakage and their directional nature, LEDs are used in a wide variety of applications. Some of the applications include recessed downlights, task lighting, traffic lights, vehicle brake lights, TVs and display cases, parking garage lighting, walkway, outdoor area lighting, refrigerated case lighting, modular lighting and many electronic device indicator lights.

Considering a large percentage of electricity output is used to power and heat/cool real estate (residential, commercial and industrial), it would be foolish to ignore the above-mentioned, time-tested and proven techniques of energy efficiency. After all, it helps us keep more money in our pockets while contributing to a cleaner environment.

Hydro Energy

Hydropower is the new and improved cousin of the old waterwheel. They both use the kinetic energy of flowing water to spin an attached element. Waterwheels converted the energy into mechanical energy by spinning an attached wheel or bucket. That mechanical energy is then used to grind grain and other physical work. Hydropower converts the kinetic energy of flowing water to electricity by using an attached generator. In a sense, it is used in the same way coal or natural gas is used to produce electricity. Both coal and natural gas are used as fuel to boil water. The resulting steam is then used to spin a turbine. The turbine is attached to an electricity-producing generator via a shaft. Hydropower uses the energy of flowing water to spin a similar generator to produce electricity.

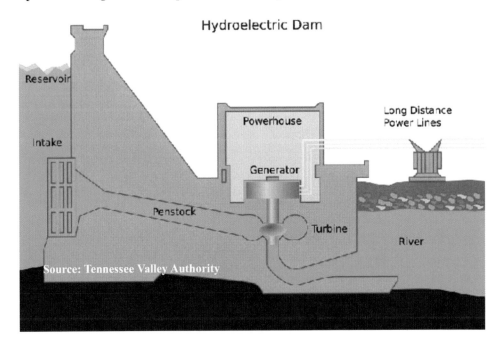

In theory, a dam is built on a large river that has a large drop in elevation. The dam stores lots of water behind it in a reservoir. Near the bottom of the dam wall is the water intake. Gravity causes water from the reservoir to fall through the penstock inside the dam. The turbine propeller at the end of the penstock is span by the moving water. The propellers are attached to an electricity producing generator via a shaft. The generated electricity is then carried to consumers by power lines that are connected to the generator. In the meantime, the water continues past the propeller through the tailrace into the river past the dam. An American icon, the Hoover Dam, located on the Colorado River on the borders of Nevada and Arizona, is an example of this kind of hydropower. In addition, hydropower facilities come in smaller dams and even dam-less, taking advantage of less powerful water flows and thus eliminating the expenses of building a dam.

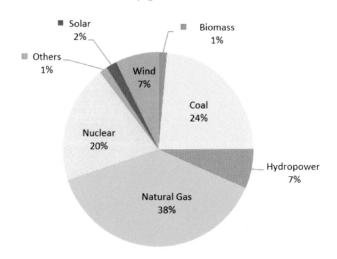

Hydropower electric share of total electricity generation in the U.S. and globally

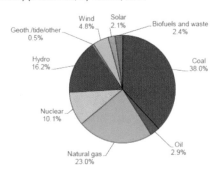

Economically, hydropower contributes a small percentage to the overall U.S. employment. In fact, the sector creates slightly over 12,000 direct jobs in the U.S. annually. A mere number compared to other renewable energy sources like solar and wind which contribute over 100,000 jobs annually. However, hydropower has the advantage of being the most reliable and one of the lowest cost of renewable energy for electricity production. Globally, hydropower generates 1100 GW. This represents about 16% of total electricity output and 50% of all renewables. While in the U.S., about 1,400 conventional, 40 pumped-storage and other forms of hydropower plants operate to generate slightly over 100 GW of power which makes up nearly 7% of total electricity output and about 22% of all renewables.

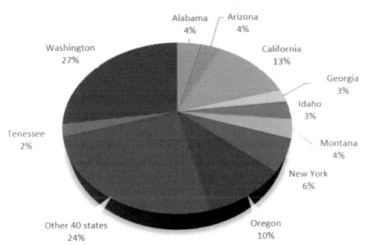

U.S. States share of hydropower electric to total hydroelectric output - 2019

Although hydroelectric contributes nearly 7% of all U.S. total electricity production and 22% of all its renewable energy output, over 50% of these outputs are concentrated in few states: Washington, California, Oregon, New York and Alabama make up 60% of all hydroelectric power in the country, while the other 45 states combine for the remaining 40% the hydroelectric power.

Hydro energy types

There are three types of commonly used hydropower plants in operation today: Impoundment, Diversion and Pumped Storage.

Impoundment: The most common type of hydroelectric power plant is an impoundment facility. An impoundment facility is typically a large hydropower system that uses a dam to store river water in a reservoir. Water released from the reservoir flows through a turbine and spins the propellers and the attached generator. In the end, the water is either allowed to continue down the river or back to the reservoir to maintain a constant reservoir level. Electricity generation capacity is heavily dependent on the amount of flowing water and the droppage height of the water. For example, large amount of flowing water will spin the turbine with large amount of force thus, generating more electricity. At the same time, the deeper the water has to flow, the more force will be impacted on the turbines, leading to more electricity generating. Thus, larger reservoirs and taller dams will produce more electricity. Below is a sample of two different kinds of Impoundment dams in the United States; **Hoover Dam** (one of the tallest dams) on the border of Arizona and Nevada and **Bonneville Dam** (one of the shortest dams) near Portland, Oregon.

Impoundment hydroelectric facilities lead to large water reservoirs
Lake Mead: an artificial lake created as a reservoir for Hoover Dam

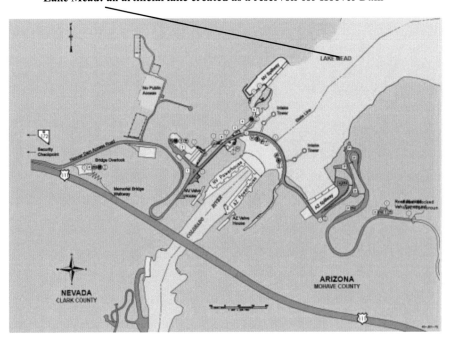

Eritrean man-made lakes
Irrigation-Drinking water supply-Power

One of the processes of building an Impoundment hydro dam is the artificial lake that is created behind the dam. The artificial lake or reservoir provides a consistent water supply to the facility as needed. In addition, the reservoir serves as irrigation and drinking water supply. The U.S., for example, has over 80,000 man-made lakes. Only 2500 of those lakes are used in hydropower. The rest are used for irrigation, drinking water supply and recreation.

The small nation of Eritrea is one of the few nations in the world that seems to understand the value of man-made lakes in developing agriculture output and achieve national food security. Since independence in 1991, the nation has built over 700 micro/small, medium and large dams and reservoirs. According to the ministry of agriculture, these dams and reservoirs have helped expand the nation's arable land and have helped promote food security.

In addition, these reservoirs contribute to the nation's water supply and help provide ample drinking water supply to citizens. In the near future, some of these dams can be used to generate electricity for the local communities or to add to the national electricity grid system. Their geographical location and small sizes make them perfect for pumped storage plants. The constant flow of water to control lake levels and for irrigation purposes make these dams ideal for diversion hydropower as well.

Some of the larger man-made Eritrean Dams/Reservoirs in operation

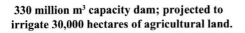

330 million m³ capacity dam; projected to irrigate 30,000 hectares of agricultural land.

Lake Mead and Hoover Dam, Arizona/Nevada Border, USA
(High Head - 726 feet: Hydraulic head - 590 feet)

Lake Mead, USA. Completed in 1935 as reservoir for Hoover dam

Water reservoirs or dams were initially made for other purposes besides power generation. They were constructed mainly for irrigation, farm ponds, water supply and flood control. The concept of generating power from hydro dams is a fairly new concept. In fact, there are over 80,000 dams in the United States, of which only about 2500 produce power.

One of the largest mad-made reservoirs is Lake Mead, where the Colorado River is backed up into and serves as a reservoir for Hoover Dam. Lake Mead was formed by the Hoover Dam project on September 30, 1935 and is the largest reservoir in the United States in terms of water capacity. The lake serves water to nearly 20 million people and large areas of farmland in Arizona, California, Nevada and parts of Mexico. Recreationally, it is a Mecca for outdoor enthusiasts who love to swim, boat, hike, cycle, camp and fish.

At its completion in 1936, the Hoover dam was the largest hydropower facility in the world. There are a total of 17 Francis turbines generating over 2 GW of electricity to millions of homes and businesses in Arizona, Nevada and Southern California.

NV Powerhouse: 8 turbines
AZ Powerhouse: 9 turbines
Facility Services: 2 Pelton turbines

Water is fed to the turbines through the four water intake towers. Two towers on the Arizona side and two towers on the Nevada side. The towers are 395 ft tall each and are made of reinforced concrete and steel. The intakes divert large volume of water to the turbines below through 30-ft diameter concrete lined pipes/penstocks.

Intake Towers

Hydro Energy

Hoover Dam Intake towers and spillways Diagram

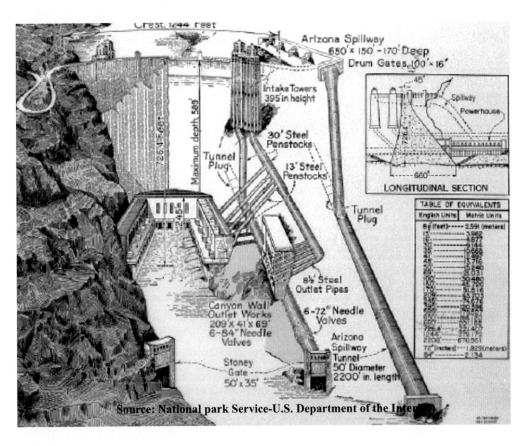

Spillways

- Concrete-lined open channels about 650 feet long, 150 feet wide and 170 feet deep on each canyon wall.
- Maximum water velocity in the spillway tunnels is about 175 feet per second or 120 miles per hour.
- Each spillway can discharge 200,000 cfs.
- If the spillways were operated at full capacity, the energy of the falling water would be about 25,000,000 horsepower. The flow over each spillway would be about the same as the flow over Niagara Falls, and the drop from the top of the raised spillway gates to the river level would be approximately three times as great.

Bonneville Dam, Oregon USA (Medium Head - 197 feet)

Bonneville Dam was completed in the late 1930s to provide hydropower (along with many other benefits, such as river navigation) to neighboring cities in the Pacific Northwest. Initial powerhouse was built with a capacity of up to ten generators producing about half a GW of power. Later on in 1981, additional powerhouse was built for an additional capacity of eight generators. Today, the facility has 20 turbines and provides around 1.2 GW of power, enough electricity to power nearly a million homes. The dam is 197 ft tall, according to the US Army Corps of Engineers. In comparison, the Hoover dam is 726 ft tall and generates about 2.0 GW of power. Today, the project servers additional benefits of flood risk management, irrigation, fish and wildlife habitat and recreation along the Columbia River.

Since the early 1930s, additional 13 dams have been built on the Columbia river. Overall, there are 60 dams in the watershed, including 14 on the Columbia, 20 on the Snake, seven on the Kootenay, seven on the Pend Oreille/Clark, two on the Flathead, eight on the Yakima and two on the Owyhee combining for a total of over 36 GW of electric power. Grand Coulee Dam, at only 550 feet high, is the largest producer of hydroelectric power in the river basin and the largest hydropower in the United States, generating about 6.8 GW with 33 turbines. Willamette Falls Dam, completed in 1888, is one of the smallest on the river basin at capacity of 15 MW at 20ft high.

The construction of the dams on the Columbia river has allowed both Oregon and Washington states to produce over 50% of their electricity, and account for over a third of all U.S. hydropower. In 2019, nearly 50% of Oregon's utility-scale electricity net generation came from hydroelectric power. Additional 11% (3.4 GW) of

Oregon's electricity net generation came from wind power generated from more than 1,900 turbines. Today, the state of Oregon produces over 60% of its electricity from renewable energy. With the addition of more wind farms, Solar power and other renewables, Oregon is pushing for 100% renewable by 2045.

(Left) inside Bonneville powerhouse (Right) older turbine

Another large dam on the Columbia river is the Grand Coulee Dam in Washington State. Grand Coulee Dam is the largest power plant by generation capacity in the United States and the seventh-largest hydropower plant in the world. The dam helps the state of Washington provide enough electricity for 4.2 million of its households annually. In fact, the state of Washington generated the most electricity from hydropower (nearly 67% of its consumption) of any state and accounted for 24% of the nation's annual utility-scale hydroelectricity generation in 2019. Since hydropower is one of the cheapest sources of electricity, states, such as Idaho, Washington and Oregon, who get most of their electricity from hydropower, have lower energy bills than the rest of the country.

Bonneville hydrodam map and navigation routes

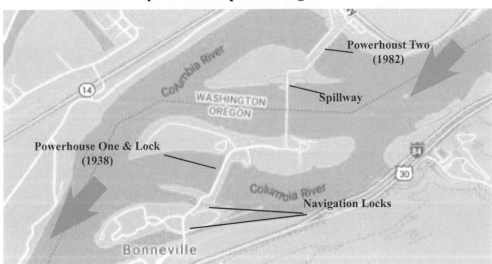

Pumped storage: Another strong hydropower region of the U.S. is the southeast along the Tennessee river valley. The region has over 29 hydropower dams generating electricity for multiple states. One of the largest power-producing plants in the area is the "Raccoon Mountain Pumped Storage plant". The plant has an intake structure of 230ft deep, allowing it to produce about 2.2 GW of power and making it one of the largest plants in the world.

Pumped storage hydropower uses the water at the upper reservoir in a similar fashion to an impoundment facility. During periods of high electrical demand, the water is released to the lower reservoir to generate electricity. Any excess energy produced is then used to pump the water from a lower elevation reservoir uphill to a higher elevation reservoir for use at a later time. This system requires at least two large bodies of water located near each other, at different elevations. The low energy density of pumped storage system requires either large flows and/or large differences in height between reservoirs. In some places, this occurs naturally while in other cases, one or both bodies of water are man-made.

If the upper reservoir collects significant rainfall or is fed by a river, then the plant may be a net energy producer in the manner of a traditional hydroelectric plant.

Unfortunately, most pumped storage facilities rely on having to pump water uphills to maintain operatable level of water at the upper reservoir. If the energy to pump the water uphills comes from renewable sources, such as solar and wind, the plant is considered a true renewalbe energy source. In some cases, however, the energy required to pump the water uphills must come from fossil fuels sources. In this case, the plant becomes simply a large capacity energy storage facility to be harnessed at will.

As of 2020, the U.S. Department of Energy Global Energy Storage Database shows a total stored energy of over 181 GW globally. China leads with 32 GW of total generating capacity while Japan and the U.S. are close 2nd and 3rd at a total generating capacity of 28 GW and 22 GW, respectively.

Raccoon Mountain pumped storage, Tennessee, USA

According to Tennessee Valley Authority (TVA), which owns the facility, Raccoon Mountain Pumped Storage works like a large storage battery. During periods of low demand, water is pumped from Nickajack Reservoir at the base of the mountain to the reservoir built at the top. The reservoir contains approximately 107 billion gallons of water covering 528 acres of water surface and takes 28 hours to fill. When demand is high, water is released via a tunnel drilled through the center of the mountain to drive generators in the mountain's underground power plant. The dam at Raccoon Mountain's upper reservoir is 230 feet high and 8,500 feet long.

The facility has four generators producing 413 MW each, for a total of 1.6 GW net capacity. Net capacity is the amount of power the plant produces on an average day, minus the electricity used by the plant itself. The facility was completed in 1978 after 8 years of construction. Today, the plant is used most days and serves as an important element for peak power generation and grid balancing in the TVA system. Globally, Raccoon Mountain plant is considered one of the largest power-producing plants. Additionally, the facility provides great recreational values, such as hiking, walking, running, road and mountain biking.

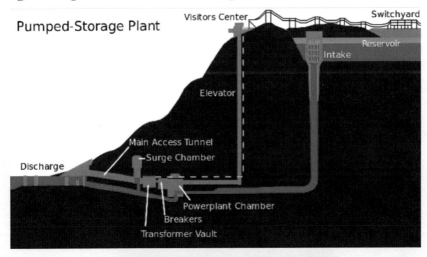

The facility has a 230-ft deep intake. The statue of liberty is 151-ft (without base) in comparison.

Because a large amount of energy is required to pump water uphills, pumped storage reservoirs are relatively small. Their small size gives them the advantage of starting up quickly. Their small sizes also make them very efficient and inexpensive.

Rank	Facility	Location	Total Capacity (MW)	Operation Start Date
1	Bath County	USA	3003	1985
2	Huizhou	China	2440	2011
3	Guangdong	China	2400	1994, 2000
4	Okutataragi	Japan	1932	1974
5	Ludington	USA	1872	1973
6	Tianhuangping	China	1836	2004
7	Tumut-3	Australia	1800	1959
8	Grand'Maison Dam	France	1800	1985
9	La Muela II	Spain	1772	2013
10	Dinorwig	UK	1728	1984
11	Raccoon Mountain	USA	1652	1978

Diversion (run-of-the-river): A diversion facility, also known as a run-of-river facility (ROR), channels a portion of the river water through a canal or penstock to a generating powerhouse before the water rejoins the main river further downstream. It can be used with a dam or dam-less system. Thus, generating electricity in a ROR system requires little water storage, known as bondage, or no water storage at all. A plant without bondage is subject to seasonal river flows. Thus, the plant will operate as an intermittent energy source.

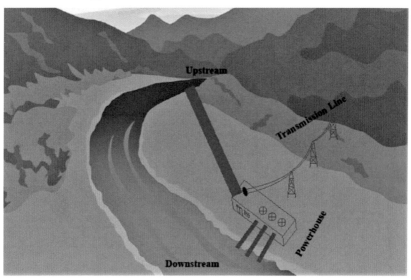

The best sites for ROR projects are areas where there is strong year-round water flow originating either from rainfall or snowpack melting and a large gravitational drop or hydrostatic head to enhance the water's energy. A greater drop in elevation means more gravitational force acts upon the water, increasing its kinetic energy.

P (power produced) = Q (water flow-meters/second) • H (hydrostatic head of water-meters) x 7.83

Disadvantages of ROR are that when the river's water levels are depleted because of drought or water extraction, the power output is reduced or becomes entirely unavailable. Basically, Q (water flow) suffers. The lack of a reservoir also puts an upper limit on the size of the run of river plant. Thus, they are only feasible on rivers with large year-round flow rates.

However, the ***advantages*** are plenty.

- The absence of water storage or reservoir makes these plants very inexpensive.
- ROR systems can be installed at existing dams, as independent generating facilities or in private systems to power small communities.
- Sometimes, plants are constructed in conjunction with river and lake water-level control and irrigation systems, such as the lake on the right.
- The lack of a major reservoir reduces the environmental footprint of run of river plants. In large hydro projects, the creation of a reservoir effects local communities as well as plant and animal life. Standing water in a reservoir can hurt overall water quality as well. With run of river systems, these impacts are not wholly avoided, but they are minimized to what is often considered a tolerable degree.

Man-made lake, Eritrea Africa: Large flows of water to control lake-level and for irrigation purposes create an ideal condition for ROR facility. However, since the reservoir is filled by rainwater, power production would be seasonal.

Typically countries that have many large hydroelectric power projects are also leaders in ROR hydropower projects. For example, China, Brazil and, Canada, possess the largest hydropower generating capacity. They are also leaders in ROR power systems. In fact, more than half of the world's ROR capacity is in China.

Chief Joseph Dam

Chief Joseph Dam is a concrete gravity run-of-the-river facility on the Columbia river, near Bridgeport, Washington. Since there is no reservoir, water flows to Chief Joseph Dam from Grand Coulee Dam up in the river basin. This is an example of an ROR facility installed at existing dams as independent generating facilities.

At a capacity of 2.6 GW, Chief Joseph Dam is one of the largest hydroelectric power producers in the United States. The single powerhouse is over a third of a mile long and holds 27 main unit penstocks to deliver water to each of the 27 house-sized Francis turbines. These 27 turbines/generators produce enough power to supply the whole Seattle metropolitan area. In addition, the facility has two small penstocks that supply water for two station service generators used to supply the power used to operate the dam.

Top ten largest hydroelectric power stations in the United States by installed capacity-2016

Rank	Facility	Capacity (MW)	State	Operating Year	Type
1	Grand Coulee	6,809	Washington	1942	Conventional
2	Bath County	3,003	Virginia	1985	Pumpmed Storage
3	Robert Moses Niagara	2,675	New York	1961	Pumpmed Storage
4	Chief Joseph	2,614	Washington	1979	Run-of-the-river
5	John Day	2,485	Oregon/Washington	1971	Run-of-the-river
6	Ludington	2,172	Michigan	1973	Pumpmed Storage
7	Hoover	2,080	Arizona/Nevada	1936	Conventional
8	The Dalles	1,813	Oregon/Washington	1957	Run-of-the-river
9	Raccoon Mountain	1,616	Tennessee	1978	Pumpmed Storage
10	Castaic	1,500	California	1973	Pumpmed Storage

Willamette Falls Dam, Oregon USA (Low Head - 40 feet)

Located in Willamette Falls, Oregon City, Oregon, Willamette Falls Dam is one of the oldest and the smallest ROR facilities in the U.S. Even though the dam has a natural waterfall of only 40 feet, it has been harvested to provide hydroelectric since 1888.

Historically, the falls served power to lumber mill, flour mill, woolen mill and paper mill since the 1840s. In 1889, Station A (later decommissioned and replaced) in Oregon City delivered power 14 miles to Portland, Oregon making it the longest distance AC power delivery at the time. It also meant that it was the oldest power plant west of the Mississippi. The facility still generates about 16 MW of electricity today.

The horseshoe-shaped dam project is located along the crest of Willamette Falls and consists of a 600-foot spillway section and a 2,300-foot dam topped with flashboards. The powerhouse (T.W. Sullivan, TWS) contains 13 turbines/generators with a total generating capacity of 16 MW.

Willamette Falls operates in a run-of-the-river mode and does not provide useable water storage or flood control. However, at only 40 feet drop, this plant is an excellent example of harvesting power from any flow of water, regardless of the volume or drop. The Willamette Falls Dam diverts water into the T.W. Sullivan powerhouse forebay on the west side of the river. The water intake for the turbines are located at the base of the powerhouse. Water diverted through the powerhouse rejoins the main river immediately below the Falls.

Mountainous countries with enormous hydropower resources are investing in run-of-river power to top up their power generation portfolios, too. Nepal, Norway, Switzerland and Austria being notable examples. In the United States, the mountainous region of the Pacific Northwest is a great location for ROR system. It is no wonder the biggest one in the country, Chief Joseph Dam, is located in this region along the Columbia River.

Sizes

Hydropower facilities range from large power plants that supply many consumers with electricity to small and micro plants that individuals operate for their own energy needs or to sell power to utilities. The Department of Energy (DOE) defines hydropower sizes as follows:

Large Hydropower: Facilities that have a capacity of more than 30 megawatts (MW).

Small Hydropower: Facilities that generate 10 MW or less of power.

Micro Hydropower: A small facility with a capacity of up to 100 kilowatts (KW) A small or micro-hydroelectric power system can produce enough electricity for a home, farm, ranch or a village.

The nuts and bolts of hydropower

Once flowing water is established, what goes on in the powerhouse is complicated and fascinating engineering work. The U.S. Army Corps of Engineers (USACE) explains the process in this way:

"A hydraulic turbine converts the energy of flowing water into mechanical energy. A hydroelectric generator converts this mechanical energy into electricity. The operation of a generator is based on the principles discovered by Faraday. He found that when a magnet is moved past a conductor, it causes electricity to flow. In a large generator, electromagnets are made by circulating direct current through loops of wire wound around stacks of magnetic steel laminations. These are called field poles and are mounted on the perimeter of the rotor. The rotor is attached to the turbine shaft, and rotates at a fixed speed. When the rotor turns, it causes the field poles (the electromagnets) to move past the conductors mounted in the stator. This, in turn, causes electricity to flow and voltage to develop at the generator output terminals."

As previously discussed, a major component of the hydropower system is the turbine. There are certain factors that need to be considered in choosing turbines, including:

- how deep the turbine must be set
- turbine efficiency
- manufacturing material and cost
- site parameters that will determine the most suitable turbine.

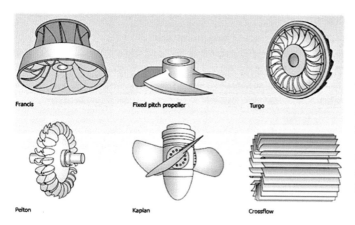

Present-day hydro turbines come in a variety of shapes. They also vary considerably in size, with 'runner' diameters ranging from as little as a third of a meter to some 20 times.

(Left) Some of the commonly used types of turbines.

There are two main types of hydro turbines used depending on the height of standing water (known as head) and the flow/volume of water.

Impulse turbine: used for high head (>1000ft) and low flow systems. Impulse turbine generally uses the velocity of the water to move the runner and discharges to atmospheric pressure. The water jet hits each bucket and pushes on the turbine's curved blades, which changes the direction of the flow. The resulting change in momentum (impulse) causes a force on the turbine blades. There is no suction on the downside of the turbine, and the water flows out the bottom of the turbine housing after hitting the runner. Examples of Impulse turbines include:

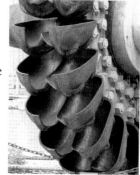

Pelton Wheel: A Pelton wheel has one or more free jets discharging water into an aerated space and impinging on the buckets of a runner. Some of these turbines have a runner diameter of up to 4 meters for heads of 200-1800 meters. They are capable of an output of up to 350 MW.

Turgo Wheel: is a variation of the Pelton and is a cast wheel whose shape generally resembles a fan blade that is closed on the outer edges. The water stream is applied on one side, goes across the blades, and exits on the other side.

Cross-Flow: A cross-flow turbine is drum-shaped and uses an elongated, rectangular-section nozzle directed against curved vanes on a cylindrically shaped runner. The cross-flow turbine allows the water to flow through the blades twice. The first pass is when the water flows from the outside of the blades to the inside; the second pass is from the inside back out.

Water Wheel: A water wheel is a machine for converting the energy of flowing or falling water into useful forms of power, often in a watermill. A water wheel consists of a wheel (usually constructed from wood or metal), with a number of blades or buckets arranged on the outside rim forming the driving car. Water wheels were still in commercial use well into the 20th century, but are no longer in common use. Water wheels were normally used for milling flour in gristmills, grinding wood into pulp for papermaking, hammering wrought iron, machining, ore crushing and pounding fiber for use in the manufacture of cloth.

Reaction turbine: is the second type of turbine used for low head (<100ft) and/or for medium head (100-1000ft) with high flow systems. Reaction turbines are acted on by water, which changes pressure as it moves through the turbine and gives up its energy. They must be encased to contain the water pressure (or suction), or they must be fully submerged in the water flow. In this system, the runner is placed directly in the water stream flowing over the blades rather than striking each individually. Examples of reaction turbines include:

Propeller: A propeller turbine generally has a runner with three to six blades in which the water contacts all of the blades constantly. The major components include the runner, scroll case, wicket gates and a draft tube.

Bulb Turbine: The turbine and generator are a sealed unit placed directly in the water stream. The term "Bulb" describes the shape of the upstream watertight casing, which contains a generator located on a horizontal axis.

Straflo: The generator is attached directly to the perimeter of the turbine.

Tube Turbine: The penstock bends just before or after the runner, allowing a straight-line connection to the generator.

Kaplan: Both the blades and the wicket gates are adjustable, allowing for a broader range of operation. With a double regulation system, Kaplan turbines provide high efficiency over a broad range of configurations. The vertical configuration of the Kaplan turbine allows for larger runner diameters (above 10 m) and increased unit power, as compared to Bulb Turbines. Some modern Kaplan turbines are also engineered with a "fish-friendly" structure to improve the survival rate of migrating species and water-lubricated bearings and water-filled hubs to prevent water pollution.

The Kaplan turbine was an evolution of the Francis turbine. Its invention allowed efficient power production in low-head applications, which was not possible with Francis turbines. The head ranges from 10 to 70 meters (33 to 230 ft) and the output ranges from 5 to 200 MW. Runner diameters are between 2 and 11 meters (6 ft to 36 ft). Turbines rotate at a constant rate, which varies from facility to facility.

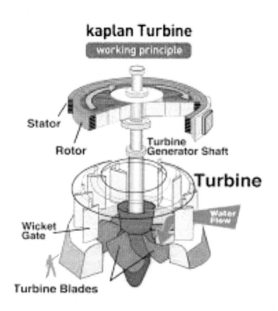

Kinetic: Kinetic energy turbines are known as free-flow turbines. They use the kinetic energy of flowing water as in rivers or streams rather than the potential energy from the head. The system is ideal for application in rivers, man-made channels, tidal waters or ocean currents. No damming or water diversion is required. The system uses the natural pathway of the water source.

Some water turbines are designed for pumped-storage hydroelectricity. They can reverse flow and operate as a pump to fill a high reservoir during off-peak electrical hours, and then revert to a water turbine for power generation during peak electrical demand. This type of turbine is usually a Francis or Dereiaz turbine in design.

Francis: A Francis turbine has a runner with fixed buckets. Water is introduced just above the runner and all around it and then falls through, causing it to spin. This is by far the most common type in present-day medium or large-scale plants with heads as low as 2m or as high as 200m. In fact, 60% of the global hydropower capacity uses Francis turbines.

Modern Francis turbines have outputs of up to 800 MW with a runner diameter of up to 10 meters and weighing over 400 tons. They can also achieve efficiencies as high as 95%. Francis turbines are very able to sustain the high mechanical stress resulting from high heads and are usually mounted with a vertical shaft to isolate water from the generator.

Deriaz/Pump turbine: Designed for pumped storage hydroelectric. They range from 30 MW to 400 MW per unit with heads up to 1,000m. Fast startup times of just 90 seconds for up to 400 MW allow for an increased number of daily starts and stops, adding flexibility and availability.

Turbine blade materials

Considering turbine blades are constantly exposed to water and extreme water pressure, they need to have high corrosion resistance and strength. Thus, most blades are made of martensitic stainless steels, which have high strength compared to austenitic stainless steels. In addition, a small percentage of chromium is added to the steel alloy to increase corrosion resistance and expand the lifespan of the blades. Martensitic stainless steel alloys also have low density and thinner sections. These qualities allow the blades to rotate more easily and lead to overall turbine efficiency. Finally, the turbines also need a higher weld quality for easier repair.

Advantages of hydropower

Hydroelectric power is much easier to obtain and more widely used than most people realize. In the U.S., all but two states (Delaware and Mississippi) use hydropower for electricity, some more than others. The state of Washington has the highest percentage of hydroelectric power generation. About 62 percent of all its electricity comes from it. Some of the reasons why hydropower is common include:

- Hydroelectric power is a domestic source of energy, allowing each state or region to produce energy independently.
- Because hydropower plants can generate power to the grid immediately, they provide essential backup power during major electricity outages or disruptions.
- Impoundment hydropower creates reservoirs that offer other benefits in addition to hydropower. They can be used for irrigation, drinking water supply, and various recreational opportunities, such as fishing, swimming and boating.
- Hydropower systems create dams that can help flood control.

Hydropower Facts (us army of corp engineers)

Hydropower is more efficient than any other form of electrical generation. It is capable of converting 90 percent of available energy into electricity. The best fossil fuel plant is only about 50 percent efficient.

Hydropower is a low-cost alternative. On average, hydropower production costs one-third that of nuclear or fossil fuel production.

Hydropower can easily respond to power needs. Hydropower dams have the ability to be turned on and off quickly. Other forms of electricity production, such as a coal-power, require a great deal of time to start or stop producing electricity.

Hydropower is a clean, reusable source of electricity. It produces no emissions and its fuel (water) can be used at each downstream dam.

Hydropower is domestic and the supply of water is continually replenished through rain and snowmelt. We are not dependent upon foreign fuel supplies and their possible interruption.

In the United States, it is now illegal to block the migration of fish, so fish ladders must be provided by dam builders.

Wind Energy

Wind energy has existed for thousands of years. The required materials and sizes to produce sizeable energy for our modern lifestyles are much more advanced. However, the technique or the concept is not new at all. Wind Energy is one of the most ancient technologies known to man. As previously mentioned, Egyptians used wind energy to propel boats along the Nile river as far back as 5,000 BC. Windmills with woven-reed blades were used for grinding grain in Persia, the Middle East and China centuries ago. Wind pumps were used to pump water since at least the 9th century in Afghanistan, Iran and Pakistan before they became widespread across the muslim world and later spread to China and India. By the 11th century, wind pumps and windmills were used extensively for food production worldwide, including in Europe, where merchants and crusaders introduced it. In Europe, particularly in the Netherlands and in the East Anglia area of Great Britain, windmills were used to drain land for agriculture.

Later on, European immigrants eventually brought wind energy technology to the Western Hemisphere, where windmills were used to grind grain, pump water and cut wood at sawmills. Ranchers and farmers installed thousands of wind pumps as they settled the western United States. With the increased electrification of rural communities in the 1930s, wind pumps and windmills started to decline. Today, we use similar wind technology as the ancient windmills and wind pumps. The only difference is that we have learned to attach generators to the turbines in order to convert the mechanical energy to electrical energy.

Wind energy technology expanded in the wake of oil shortages and climate change concerns. In response to the oil embargo of 1970 and the Iranian revolution of 1979, the United States, along with other nations, started investing in the development of alternative energy sources. The U.S. federal government provided research and development funding to reduce the cost of wind turbines and offered tax and investment incentives for wind power projects. By the early 1980s, thousands of wind turbines were installed in California, largely supported by federal and state policies that encouraged renewable energy sources.

A few years later, the federal government established incentives to use renewable energy sources in response to renewed concerns over climate change and fluctuating fuel prices. Congress passed the Production Tax Credits (PTC) in 1992 and Investment Tax Credit (ITC) in 2006 to incentivize the production and distribution of electricity from renewable sources. In addition, state governments enacted new requirements for electricity generation from renewable sources. Soon, electric power marketers and utilities began to offer electricity generated from wind and other renewable energy sources, called green power, to their customers. The tax credits for wind and solar anchored by advances in technology contributed to a remarkable drop in the cost of renewables projects. In the past decade alone, wind power costs have declined by nearly 70%. As a result, U.S. total annual Electricity generation from wind power increased from about 6 billion kilowatt-hours (kWh) in 2000 to about 300 billion kWh in 2019, contributing to nearly 7.3% of total U.S. utility-scale electricity generation and 24% of all renewable source consumed in the United States.

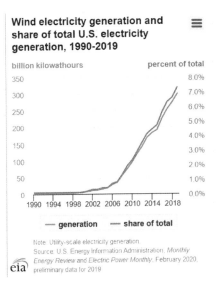

Most importantly, wind power projects were very dispersed across the U.S. with over 42 states involved in some sort of wind power projects. The five states with the most electricity generation from wind in 2019 were Texas, Oklahoma, Iowa, Kansas and California. These states combined produced nearly 60% of total U.S. wind electricity generation in 2019. Table 3 below illustrates each states wind power generation as of 2019.

Globally, world wind electricity generation has also increased substantially in recent years. In 1990, 16 countries generated a total of about 3.6 billion kWh of wind electricity. In 2000, 49 countries generated about 31 billion kWh, and in 2017, 129 countries generated about 1,129 billion kWh of wind electricity.

Table 3: Top 5 states for wind powered electricity generation in 2019

Rank	State	Electricity generation capacity (MW)	Highlights
1	Texas	24,899	Top Wind Farm: Roscoe wind farm with 634 turbines. Enough to power nearly 6 million homes
2	Iowa	8,422	4637 turbines contribute to 37% of state's electricity
3	Oklahoma	8,072	Generates 31.9% of its electricity from wind power
4	California	5,885	Alta wind energy center: Largest wind farm in the U.S. and 2nd largest in the world. Top Solar electricity producer in the U.S. Commited to 100% renewable energy by 2045
5	Kansas	5,653	3000 turbines. Commited to 20% electricity generation from wind power by 2020

In addition to contributing to cheaper and cleaner electricity, wind energy has also brought in direct investments and created new job opportunities to the U.S. economy. According to the American Wind Energy Association (AWEA), wind energy, supported by U.S. tax incentives and credits of the past decade, has created:

- 114,000 jobs and generated more than $143 billion in private investment across all 50 states.
- Contributed more than $760 million in local and state taxes.
- Resulted in nearly $290 million in lease payments to farmers annually.

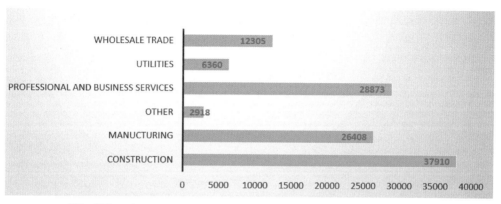

Wind Electric Power Generation jobs - Employment by Industry Sector

Wind energy is also creating new opportunities in factory towns across America. Over 530 factories across 43 states build wind energy-related parts. The above chart illustrates the total employment numbers in the electricity generation sector. The wind energy sector employs the second most of the renewable energy at 114K jobs in the U.S. and employs over 11 million people globally. Most importantly, since most of the jobs are in the construction and utility sectors, the jobs will remain locally.

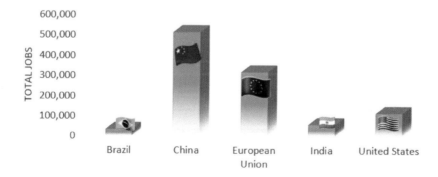

Understanding wind energy

Wind energy, at least for the purpose of discussion in this book, is the process of creating electricity using the power of the wind and wind turbines. The process crosses three energy conversions: The kinetic energy of the wind spins the blades. The blades spin the attached internal shaft creating mechanical energy. Finally, the internal shaft spins an attached generator, which produces electrical energy or electricity. Electrical wires and cables then transport the generated electricity to the grid for distribution. Wind turbines start generating electricity when wind speeds reach 6 to 9 miles per hour (mph). This starting wind speed is known as the cut-in speed. As wind speeds increase, so does electricity production. However, a wind speed of about 55 mph and above is considered too dangerous, and the turbines are shut down to prevent equipment failure. Modern wind turbines have an average life of over 25 years and can generate usable amounts of electricity over 90 percent of the time.

The process of converting wind to electricity (illustrated below) requires massive initial investment in material, of which the the tower, the blade and the nacelle are the biggest factors in determining the amount of electricity a wind turbine can generate. Bigger blades on a taller tower can capture more wind to run a bigger generator. The largest wind turbines in operation today have electricity generating capacities of up to 10 megawatts (MW).

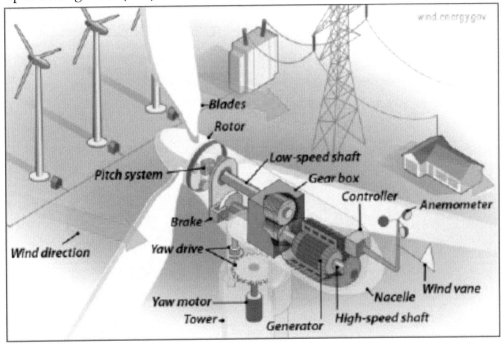

Wind Turbine

An electricity generating structure. Converts the kinetic energy of the wind to electricity using many components including blades and generators.

Blades: shaped like an airplane wing. Spun by wind to initiate the electricity generating process

Nacelle: a large box where the turbine, shafts and the generator are stored

Tower: a structure made up of steel and concrete. Supports the blades and nacelle.

Transformers: housed on the floor of the tower. They up the voltage for long range transport where voltage levels are adjusted to match with the grid.

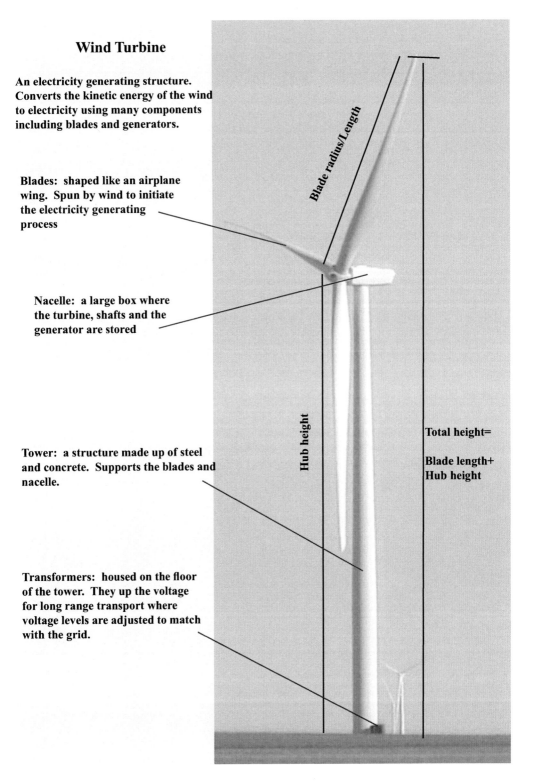

Blade radius/Length

Hub height

Total height= Blade length+ Hub height

Taller towers lead to more electricity generation
Towers today are taller and cover larger land acreages than ever before

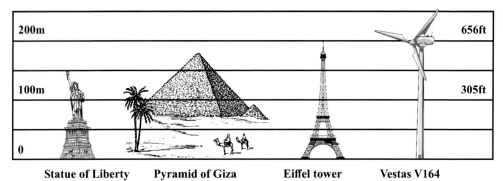

Towers: The towers are the backbone/support of the whole structure. They support the blades and nacelle structure, which houses the shaft, gearbox, generator and controls. The tower also houses ladders or elevators for crew members to climb up and down as well as electrical cables to transport electricity from the generator to the transformer down below.

Because wind speed increases with height, taller towers enable turbines to capture more energy and generate more electricity. For example, the widely used GE 1.5-megawatt turbine model, consists of 116-ft blades atop a 212-ft tower for a total height of 328 feet. The blades sweep a vertical airspace of just under an acre. The 1.8-megawatt Vestas V90 from Denmark has 148-ft blades (sweeping more than 1.5 acres) on a 262-ft tower, totaling 410 feet. Another model being seen more in the U.S. is the 2-megawatt Gamesa G87 from Spain, with 143-ft blades (just under 1.5 acres) on a 256-ft tower, totaling 399 feet. Many existing models and new ones being introduced reach well over 600 feet in total height. Table WE01 lists some of

Table WE01: Sample of some of the smaller and larger turbines in operation

Turbine Model	Capacity (MW)	Blade Length m, ft	Hub Height m, ft	Total Height m, ft	RPM range	Rated wind speed m/s, mph
Bonus (Siemens)	1.3	31, 102,	68, 223	99, 325	13-19	14, 31
GE 1.5s	1.5	35.25, 116	64.7, 212	99.95, 328	11-22	12, 27
Vestas V112	3	56, 184	84, 276	136, 459	6-17	12, 27
Enercon E-126	7.6	63.5, 208	135, 443	198.5, 651	5-11	
Vestas V164	9.5					

the commonly used tower structures.

Towers must also be extremely strong to support the equipment. For example, In the GE 1.5-megawatt model, the nacelle alone weighs more than 56 tons, the blade assembly weighs more than 36 tons and the tower itself weighs about 71 tons totaling a weight of 164 tons. The corresponding weights for the Vestas V90 are 75, 40, and 152, total 267 tons; and for the Gamesa G87, 72, 42, and 220, totaling 334 tons. Therefore, the supporting structure, which is the tower, must be extremely strong and is normally made of steel or concrete.

The three main types of towers are:

Tubular Steel Tower

The steel wind turbine tower is the most commonly seen tower type in the world. The steel tower is made in sections of around 65-130 ft (20-40m). The sections are connected with wind tower flanges. The flanges are then bolted together. All steel wind towers are in taper shape, meaning the diameter of each section decreases as the tower height increases. Because of their closed structure, the internal parts are well protected from the weather. In addition, antirust paint and coating are sprayed manually to protect the towers from corrosion, adding to the overall cost. Their large sizes also create challenges for transporting them to job sites.

Concrete Tower

Similar to the steel towers, the precast concrete wind turbine towers are also manufactured in sections. All the sections are transported to the site and then assembled. These concrete structures have a large carrying capacity allowing them to bear larger blades and nacelle. Since there are no fastening parts, the repair and maintenance cost is reduced largely.

Lattice Tower

Lattice wind turbine towers are made from hundreds of steel materials. They look just like the traditional telecommunication towers. Many of these towers are made by companies who manufacture the tradional telecommunication or electric towers since they have the experience of making these towers. Since these towers are made in smaller steel sections, transportation is easier and the cost is lower. However, cabling and other supporting mechanism need extra protection since they are exposed to the environment unlike the tubular steel or concrete towers.

Wind turbine towers are also self-supporting (free-standing) or guyed. Most towers are self-supporting. However, there are a few exceptions depending on the site location and terrain.

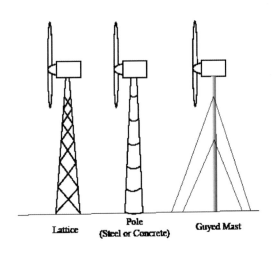

Blades: The size of blades varies widely, spanning up to 150 feet wide and resemble the wings of an airplane. Wind flows over the blades creating lift (similar to the effect on airplane wings), which causes the blades to turn. Aerodynamic properties are crucial in determining how well wind turbine blades can extract energy from the wind and efficiently produce wind power.

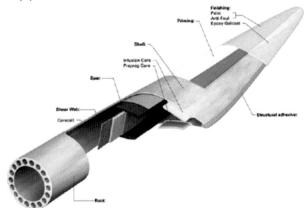

Blades are usually fabricated from thin-walled fiber composite materials, with fiberglass-reinforced polyester or epoxy. Carbon fiber or aramid (Kevlar) is also used as reinforcement material. Fiber composite materials are high in strength and stiffness, lightweight, low density and have superior fatigue properties. Nowadays, the possible use of wood compounds, such as wood-epoxy or wood-fiber-epoxy, is being investigated. Below are different types of fiber products used in manufacturing blades today.

Glass fibers: The stiffness of composites is determined by the stiffness of fibers and their volume content. Typically, E-glass fibers (i.e., borosilicate glass called "electric glass" or "E-glass" for its high electric resistance) are used as main reinforcement in the composites. By increasing the volume content of fibers in the composites, the stiffness, tensile and compression strength increase proportionally. Typically, the glass/epoxy composites for wind blades contain up to 75 weight % glass.

Carbon fibers: Carbon fiber has more tensile strength, higher stiffness and lower density than glass fiber. Carbon fibers are also 20% to 30% lighter than glass fibers, leading to thinner, stiffer and lighter blades. However, because carbon fiber is more expensive, glass fiber is still dominant. Carbon fiber reinforced composites also have the disadvantage of being too sensitive to the fiber misalignment and waviness, which lead to reduction of strength.

Aramid and basalt fibers: These are an alternative for non-glass, high strength fibers. Aramid (aromatic polyamide) fibers demonstrate high mechanical strength and are tough and damage tolerant. They are 30% stronger, 15–20% stiffer and 8–10% lighter than E-glass, and cheaper than the carbon fibers. Unfortunately, they have low compressive strength and low adhesion to polymer resins. They also absorb moisture and degrade due to ultraviolet radiation. Currently, basalt fibers are used in small wind turbines and as hybrids with carbon fibers.

Hybrid composites: Hybrid reinforcements (E-glass/carbon, E-glass/aramid, etc.) are an alternative to pure glass or pure carbon reinforcements. Hybrids lead to much lower blade weight and unfortunately, to higher cost. Thus, their use is limited.

Natural fibers: Although not common, natural fibers can be used as well, especially in residential or small scale wind projects. Natural fibers, such as sisal, flax, hemp and jute, are low cost, readily available and environmentally friendly. Unfortunately, they have high moisture uptake and low thermal stability leading to quality inconsistency. Another promising natural fiber is bamboo. Bamboo has high strength and durability and is also broadly available. Locally available timber is also promising material for low cost and reliable wooden blades.

Blade maintenance

Blades are the most vulnerable parts of a wind turbine. They are consistently exposed to extreme weather conditions, including lightning. Lightning strikes are an expected reality of a wind turbine. It is common to observe scorching damage and cracking of blades, as well as spar rupture, separation and surface tearing in more extreme cases. Thus, all blades have a lightning protection system to reduce the effect of such strikes.

In addition, harsh weather conditions also cause significant blade damages. For example, icing on the surface of the blades under extremely low temperature weather conditions will degrade and even stop the operation of the turbine as was seen in the Texas storm of 2021. The aerodynamics of the blade will suffer and energy generation will be reduced. Also, the additional ice weight leads to unbalanced load distribution, which will lead to structural fatigue.

Airborne particulates also cause significant damage, especially around the tip of the blades, where velocities are higher. These damages lead to a rough surface which will degrade the aerodynamic performance of the blade and reduce power production. Therefore, continuous maintenance is a must for optimal results.

Nacelle: A nacelle is a cover housing that sits atop the tower and houses all of the electricity-generating components. The housing frame is made of front and rear frame parts. The front or main frame of the nacelle is generally made of cast steel and holds the yaw system, gearbox and the drive shaft that is connected to the blades via the hub. The rear frame is constructed of formed and welded steel and houses the generator, transformer and electrical control cabinets.

Once the yaw system passes its rotational test and its motors are installed and pass their functional tests, the two halves of the frame are joined by heavy bolts and spring pins. The entire assembly is attached by brackets to the bottom half of the nacelle's fiberglass cover.

Then the main shaft and gearbox unit along with the generator assembly are lifted into the nacelle using a gantry crane and bracketed to the gondola. Several OEMs install the transformer inside the nacelle at this point, but most install the transformer at the base of the tower.

With everything in place, fiberglass upper housing is installed to cover and protect components inside. Finished nacelles are then moved out of the factory and shipped by truck or rail to wind farms to be lifted onto towers.

Below are some of the smallest and some of the biggest Nacelles in operation around the world

Nacelle - System and Components

The functional groups and systems within a nacelle of a modern wind turbine include: the yaw system, the mechanical drive train, and the electrical systems and cabinets

Yaw System

The Yaw System is the component that is responsible for the orientation of the wind turbine rotor towards the wind. It allows the turbine to head itself into the wind direction, but also keeps itself locked in that position when wind direction is stable. There are two techniques used; both are based on a bearing, brakes, drives, and positioning system:

Yaw Roller Bearing System/Passive yaw system: This technique utilizes the wind force to adjust the orientation of the rotor into the wind. It comprises a roller bearing connection between the tower and the nacelle. A tail fin/wind vane is mounted on the nacelle to help the rotor turn into the wind by exerting a "corrective" torque to the nacelle. Simply, the power of the wind is responsible for the rotor rotation and the nacelle orientation. This system is commonly used for small/medium size wind turbines since it offers a low cost and reliable solution.

Yaw Slide Bearing System/Active yaw system: This technique is based on automatic signals from wind direction sensors. This system is considered to be state of the art for all modern medium and large sized wind turbines, including offshore turbines.

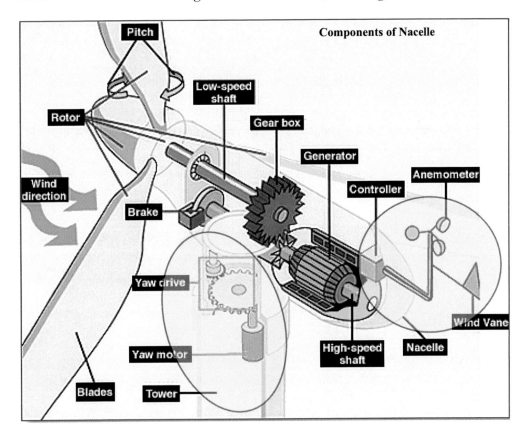

Brake: Stops the rotor mechanically, electrically or hydraulically in emergencies or if wind speed is unsafe.

Pitch: Turns (or pitches) blades out of the wind to control the rotor speed and to keep the rotor from turning in winds that are too high or too low to produce electricity.

Anemometer: A measuring device that sits atop of the Nacelle. It measures the wind speed and direction and transmits the data to a controller.

Controller: Starts up the machine at a cut-in wind speed and shuts off the machine at about 55 mph to prevent turbine damages at higher wind speeds. Also, it uses the data from Anemometer to rotate rotor/blades to face maximum wind direction.

Mechanical drive train

The mechanical drivetrain is the "powerhouse" of a wind turbine. It contains the gearbox and the generator, which are responsible for converting the rotation of the blades into electricity.

The gearbox speeds up the slow rotation of the rotor from around 5-15 rotations per minute (rpm) to higher speeds of 1,000–1,800 rpm needed to generate electricity. This is called *Gearbox Drive*.

The other alternative is *Direct Drive*. Direct drive generators can produce electricity at much lower speeds. They are directly attached to the rotor and do not require a gearbox. However, their makeup requires heavy, rare earth materials such as neodymium and dysprosium, which makes them expensive. These are mainly used in larger offshore turbines.

Rotor: The blades and hub together form the rotor, which is attached to a low-speed shaft.

Low-speed shaft: a metal rod connected to the blades on one end and a gearbox on the other end. The shaft is span by rotating blades.

Gearbox: Connects the low-speed shaft to the high-speed shaft to increases the rotational speeds to a required rpm of about 1,000-1,800. The gearbox is one of the more costly and heavily weighted parts of the wind turbine. There is ongoing research and development to reduce the cost of previously discussed "Direct Drive" generators and eliminate gearboxes from the equation.

High-speed shaft: Drives the generator.

Generator: The part responsible for producing electricity.

Electrical Systems and Cabinets

The Electrical Systems and Cabinets convert the electricity of the generator into a suitable format to match the Electric Grid, frequency AC/DC.

Transformers: Finally, all the electricity generated by the wind turbine and the generator is carried down to a transmission substation via electrical wires. The electricity is then converted into extremely high voltage, between 155,000 and 765,000 volts, for long-distance transmission on the grid. The grid comprises a series of power lines that connect the power sources to demand centers.

Wind turbine types

There are two basic types of wind turbines: Horizontal-axis turbines and Vertical-axis turbines. Most wind turbines manufactured today are horizontal-axis with three blades

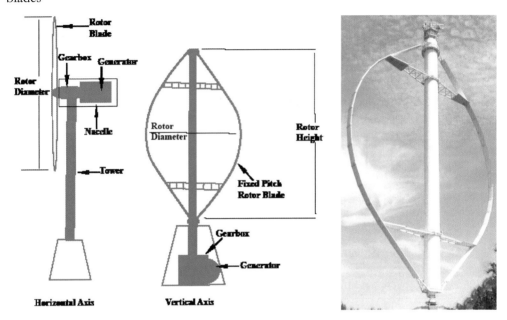

Horizontal-axis turbines have three blades that resemble propeller airplane engines. The largest horizontal-axis turbines are as tall as 20-story buildings and have blades more than 100 feet long. Longer blades on taller towers generate more electricity. Nearly all of the wind turbines currently in use are horizontal-axis turbines.

Vertical-axis turbines look like egg beaters. The blades are attached to the top and the bottom of a vertical rotor and a look of a giant two-blade egg-beater. Some versions of the vertical-axis turbine are 100 feet tall and 50 feet wide. Very few vertical-axis wind turbines are in use today because they do not perform as well as horizontal-axis turbines.

Site allocation

How does one decide where to install a wind farm?

Wind power plants require careful planning that extends beyond merely erecting wind turbines in a windy area. A detailed study of the site, based on topology and wind blow characteristics (how fast or slow and how often wind blows), must be studied ahead of time. After all, wind speeds generally change throughout the day and from season to season. For example, in parts of California, where numerous wind turbines are located, the wind blows more frequently from April through October, and the wind is usually strongest in the afternoon. In Montana, strong winter winds channeled through Rocky Mountain valleys create more intense winds during the winter.

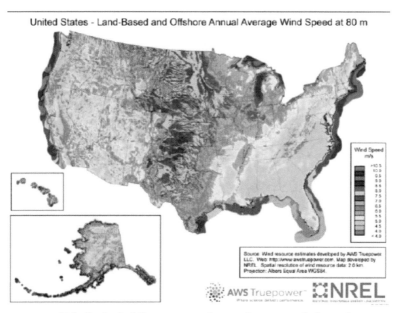

U.S. Geological Survey map of annual average wind speed

The right places for wind turbines are where the annual average wind speed is at least 9 mph for small wind turbines and 13 mph for utility-scale turbines. At these low required speeds, every region has the potential to produce wind energy. However, favorable sites are areas with consistent wind resources such as tops of smooth and rounded hills, open plains, oceans/seas and mountain gaps that funnel and intensify wind. In addition, since higher elevation produces more energy, wind turbines are placed on towers that range from about 500 feet to as much as 900 feet tall.

Wind farms

Large turbines that are grouped over a large area are called wind farms. Wind farms provide power to large scale utility companies' electricity grids. Wind farms can either be onshore (on land), like most projects today or offshore (on oceans/seas), which is gaining acceptance rapidly.

Onshore wind farms: These are wind turbines installed on land. Because onshore wind farms are installed on land, costs can be much lower than the other option. Still, installation is no easy walk. It is extremely challenging and requires heavy duty machinery due to the turbine size and weight. Below is a pictorial illustration of an installation process for a single turbine.

Wind turbine installation
(Left) Tower base/foundation preparation and erecting the tower in sections
(Middle)installation of rotor hub and nacelle (Right)attaching the blades

Image, installing hundreds of turbines as is the case with wind farms. They require a large acreage of land space despite land limitations due to terrain challenges and urban expansion. To provide an uninterrupted flow of air ensuring power generation, onshore wind turbines require a minimum distance of 150m from any obstructions and a turbine-to-turbine separation of 7 times the diameter of the rotor.

For example, the largest onshore wind farm in the U.S. (Alta wind farm, California) has about 600 turbines generating over 1.55 GW of electricity. The farm was constructed in 2010 at a cost of $2.9 Billion dollars and covers an area of 3,200 acres (5 square miles, mi^2) of land.

Fortunately, energy generated from onshore wind farms can easily be added to the grid. The maintenance cost is quite low as well. In some cases, investment payback can be as fast as two years. For this reason, most wind farms in operation today are onshore. Onshore wind energy continues to expand and global capacity from onshore wind energy is projected to reach nearly 750 GW by 2022. Today, onshore wind energy is the 2nd cheapest source of energy behind solar energy at a cost of below 5 cents per kWh. As a reminder, the fossil fuels and nuclear energy fueled energy cost in the U.S. today is at about 12 cents per kWh for residential and 7 cents per kwh for industrial.

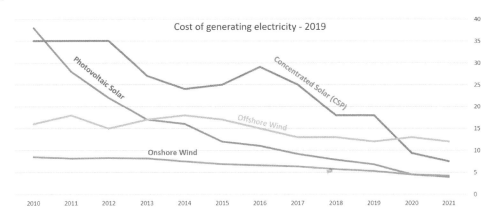

The U.S. accounts for five of the top ten onshore wind farms in the world. The largest five wind farms in the U.S. are as follows:

1. Alta Wind Energy Centre (3rd largest in the world)
Located in Tehachapi, Kern County, California, the Alta Wind Energy Centre is the biggest wind farm in the US. With a combined installed capacity of about 1.55 GW, it has about 600 turbines and covers over 5 mi^2 land area. It was commissioned in 2010 at a cost of $2.9 billion dollars.

2. Los Vientos Wind Farm
With an installed capacity of 910 MW, the Los Vientos Wind Farm in Texas was completed in five phases starting in 2012 and ending in 2016. The combined five sections of the farm have about 426 installed turbines.

3. Shepherds Flat Wind Farm
Located near Arlington in Eastern Oregon, the Shepherds Flat Wind Farm is the third largest wind farm in the U.S. with installed capacity of 845 MW. The wind farm is spread over 30 square miles of land area. Construction on the wind farm started in 2009, with an estimated cost of $2 billion dollars.

The Shepherds Flat Wind Farm began commercial operations in 2012. It consists of 338 GE2.5XL turbines, each with a rated capacity of 2.5MW.

4. Roscoe Wind Farm

The 781 MW Roscoe Wind Farm near Abilene, Texas, is the fourth biggest wind farm in the U.S. The wind farm was built in four phases between 2007 and 2009 for a total of 627 wind turbines.

5. Horse Hollow Wind Energy Centre

Located in Taylor and Nolan County, Texas, the Horse Hollow Wind Energy Centre has an installed capacity of 735.5 MW. The wind farm was commissioned in four phases during 2005 and 2006. At the time of its completion in 2006, it was the largest wind farm in the world. The farm has about 421 wind turbines generating electricity for nearly 180,000 Texan households.

Jiuquan Wind Power Base/Gansu Wind Farm, China

For comparison purposes, the largest wind farm in the world, Jiuquan Wind Power Base in China, has installed capacity of about 8 GW. The wind farm construction began in 2009. When completed, it is projected to have 7,000 wind turbines for a total installed capacity of 20 GW at a cost of nearly $17 billion.

Onshore wind farms are a great way of utilizing desert and wastelands

Offshore wind farms: In some cases, wind towers are built offshore to take advantage of the 20% higher wind flow generated in the oceans. Offshore wind farms have a higher potential for power generation due to faster and more constant winds and thus, providing energy at a more stable rate. Additionally, offshore wind farms have the advantage of large open spaces for installation and no noise pollution that is an issue with onshore farms. As a result, offshore wind turbines can be twice or three times larger than onshore options. Their larger sizes allow higher power generation and more efficiency for the cost of materials.

Although the operation is similar to onshore wind power, the installation is much more demanding considering the challenges of building tower foundations on ocean floors. Offshore wind farms also require platforms, underwater cables and interconnection, and other factors that increase installation costs by nearly 20%.

Despite the installation challenges, the advantages of limitless offshore space for turbine installations, stable wind patterns, not interfering with people's livelihoods, as well as rapidly declining costs, are far more beneficial. As a result, offshore wind power is projected to grow at a rate of 16% between 2019 and 2030, reaching a cumulative capacity of 142 GW by 2030, compared to the 23 GW reached at the end of 2018.

Globally, both the European Union and China are making great strides in expanding offshore wind farms. Unfortunately, there is no offshore wind farm to speak of for the United States. The largest offshore wind farm in the U.S. is Block Island Wind Farm, located off the coast of Rhode Island in the Atlantic Ocean. The farm has five turbines installed in 2016 for a total capacity of 30 MW. Globally, the United Kindom (U.K.) has four of the top five largest offshore wind farms in the world.

1. Hornsea

The Hornsea-1 offshore wind farm, with an operating capacity of 1.2 GW is currently the worlds' largest offshore wind farm in the world. The wind farm consists 174 turbines covering an area of 157 mi^2 (407 km^2) of North Sea off the coast of the U.K. The project is the first offshore wind farm to generate over 1 GW of electricity. According to the official project website, the farm generates enough power to meet the energy needs of over a million British households. Hornesea-1 is the first of planned four phases project for a total capacity of 6 GW electricity when completed.

2. Walney Extension

Walney Extension of England's Walney Island is the second-largest offshore wind energy farm in the world with an operational capacity of 659 MW. Operation began in 2018 with 87 turbines, enough to generate clean electricity to power 600,000 homes.

3. London Array

Third in the list is the London Array offshore wind farm with a capacity of 630 MW located off the coast of the U.K. The plant is commissioned in April 2013 with 175 turbines.

4. Gemini Wind

The only top five project located outside the U.K. waters is the Gemini wind farm in the Dutch part of the North Sea. With a generational capacity of 600 MW, the project generates electricity for approximately 785,000 households. The plant was officially commissioned in 2017 and featured 150 turbines.

5. Beatrice

The final project on the list and one of the newest is the Beatrice offshore wind farm off the coast of Scotland. The project was commissioned in 2019 with a capacity of 588 MW. The farm features 84 wind turbines and generates enough electricity to over 450,000 homes annually.

Upcoming Project

The world's largest offshore wind project, Dogger Bank Farm, in the North Sea off the coast of the U.K. is to be constructed in three phases for a total 3.6 GW capacity. According to the project website, the project will generate electricity for nearly 6 million homes in the U.K. or 5% of the U.K. electricity when completed in 2026. The project will use the world's largest, 13 MW General Electric Haliade-X, turbines. The turbines stand more than 850 feet tall and are designed to generate 45% more power than current best-in-class offshore turbines. According to GE, the turbine's blades measure longer than a soccer field (351 feet) and a single rotation can supply enough electricity to power the average British household for two days.

Small scale wind power

It is obvious that larger wind farms are more efficient and effective. However, building increasingly large turbines requires greater financial investment and resources.

The wind farms themselves require an ever increasing open spaces under the right wind conditions. As a solution to these issues, small scale wind turbines are emerging. These small turbines can be deployed in a greater range of locations and allow individuals to contribute to cleaner and renewable energy production.

Small scale turbines operate at a lower elevation of 30 ft to 140 ft compared to utility scale turbines that operate at 300 ft to 900 ft and typically need wind speeds of 9 mph. Because wind speed is slower and inconsistent at lower elevation, these turbines have small electricity generating capacity. Thus, their use is limited to residential homes, businesses, rural or farms areas. Their reduced cost and versatility also allow them to be used on-site, rather than transmitting energy over the electric grid from central power plants or wind farms.

Small scale wind power utilizes two different types of turbines:

Horizontal axis wind turbine (HAWT): Recent advances have improved many of the features of these turbines. Newer models can generate more power with fewer rotations per minute (RPMs) making them more efficient. Some even have curved blades to help reduce noise level. Additionally, new models use rare earth magnets in their generators, allowing for smaller, lighter generators. The rotors are also equipped with brakes or pitched blades to protect the turbines against damage from high winds.

Vertical axis wind turbine (VAWT): turbines are designed with blades that rotate around a vertical shaft. These turbines operate at lower wind speeds thus, can be placed at a lower elevation than HAWTs. The shape of their blades also allows them to generate power from wind blowing in any direction, including vertically. VAWTs also have smaller space requirements and can be placed closer together than HAWTs. Since the generators are located at ground level at the bottom of the shaft, maintenance is also much easier.

However, one disadvantage for these turbines is that they generate less power than their counterparts HAWTs because they operate at lower wind speeds. Also, the lack of a brake system makes the blades more vulnerable to damages by high winds.

Basically, HAWTs are far more common due to their superior efficiency and generation capabilities. However, VAWTs may be preferred in locations with less available space or where wind speed and direction are inconsistent.

Small scale wind power can be used in an off-grid or on-grid system. An **Off-grid** system is where wind turbines are not connected to the electrical transmission grid. These are generally installed in areas far from the grid and/or where connecting to the grid would is expensive. Off-grid systems use direct current (DC) to power devices in remote locations, such as telecommunications equipment and water pumps in rural areas. These systems can also use battery storage to provide backup power when the wind is not blowing.

The alternative is to use small scale wind power in an **On-grid** system where wind turbines are connected to the grid. Many newer turbines designed for on-grid system have built-in inverters to convert the electricity to Alternating Current (AC) for compatibility with household appliances. An added benefit of the on-grid system is that any excess electricity generated can be put back to the grid and sold to the utility company for profit.

A modern building in Portland, OR, with four small wind turbines (HAWT) on rooftop. Power from these small scale turbines contributes to the overall electricity requirement for the building.

Active Solar Energy

According to the National Renewable Energy Laboratory (NREL), the sun produces enough energy every second to cover the earth's needs for 500,000 years. Of that energy, enough power reaches the earth in one hour to power all of society for a year. The sun is the ultimate source of energy. It is the source of fossil fuels we have come to depend on. It is also the source of renewable energy we are learning to rely on. As previously discussed, we have used the sun's energy in passive solar design for thousands of years. Recently, we have learned to use the sun's energy in much more effective ways to generate heat and electricity for our homes and businesses. This recently gained knowledge is called "Active Solar Energy".

Active Solar Energy is radiant light and heat from the sun harnessed into clean energy to power and heat residential, commercial and industrial energy demands. This renewable energy source, first used commercially in the mid-1800s, is growing rapidly as the demand for cleaner and sustainable energy has increased worldwide in both developed and developing nations.

Active solar energy is harnessed using two different techniques: Photovoltaic system (PV) and Thermal energy.

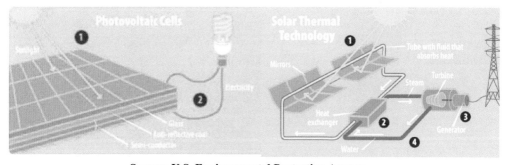

Source: U.S. Environmental Protection Agency

A photovoltaic system (PV) or commonly known as solar power, is the process of converting the sun's light directly into electricity with the help of a solar panel and additional hardware. Thermal energy, on the other hand, requires using the sun's energy to heat water or other fluids and using the steam generated from the fluid to propel a generator.

According to the Solar Energy Industries Association (SEIA), the U.S. had 64.2 GW of installed solar power at the end of 2018, enough to power 12.3 million homes. This is a huge improvement compared to 10 GW of solar power capacity ten years ago. Today, solar power accounts for 2% of total U.S. electricity generation and has ranked either first or second in capacity added to the entire U.S. electric grid system every year since 2013.

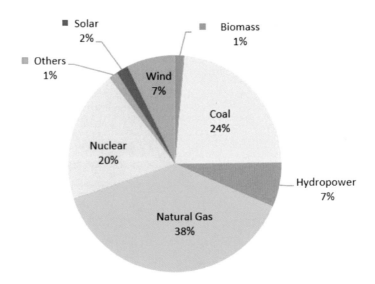

Globally, solar power accounts for only 2% of all electricity generated. Many countries and territories have already installed significant solar power capacity into their electrical grids to supplement or provide an alternative to conventional energy sources. The worldwide growth of photovoltaics varies enormously by country. By the end of 2019, a cumulative amount of 629 GW of solar power was installed throughout the world, with China accounting for one-third of it at 208 GW,

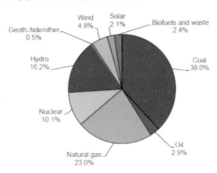

followed by the U.S. and India. As of 2020, there were at least 37 countries worldwide with a cumulative PV capacity of more than one GW.

Leading the way to this rapid growth is technological advances that have improved solar panel efficiency from 5% in the 1950s to nearly 30% today. Favorable government policies and incentives have contributed greatly to the advancement. For example, the U.S. Federal program, known as "Investment tax credit for solar power (ITC)", was enacted in 2005 and extended several times to 2021. The program allowed for tax credits and helped reduce the cost of solar power installation and contributed to an astonishing 70% decline in solar energy cost in the last ten years in the United States. The cost reduction has allowed more businesses and communities to take advantage of solar power. According to the SEIA, the average-sized residential system cost has dropped from a pre-incentive price of $40,000 in 2010 to roughly $20,000 today. The cost of generating electricity with solar power has fallen significantly from $0.38/kWh in 2010 to below $0.05/kWh today, making solar power the cheapest source of electricity in the U.S.

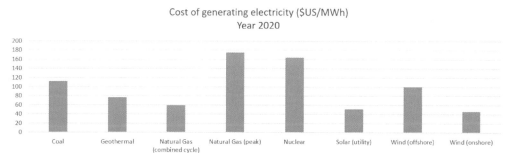

The increasing competitiveness of solar power against other technologies has allowed it to quickly increase its total U.S. and global electricity generation share from just 0.1% in 2010 to nearly 2% today

Economically, the solar power industry has made a significant contribution to the labor force as well. According to U.S. Energy and Employment Report (USEER), the solar energy sector employs over 345,000 workers, more than the coal, oil and natural gas industries combined. The solar photovoltaic installer is also the fastest-growing job in America today.

As impressive as the number of new jobs is, more stunning is that all of these jobs are spread out over nearly all 50 U.S. states. As previously mentioned, renewable energy allows each region to develop an energy portfolio mix based on its natural climate. Thus, it is no surprise to see regions with the hottest climate lead in solar power consumption and employment opportunities.

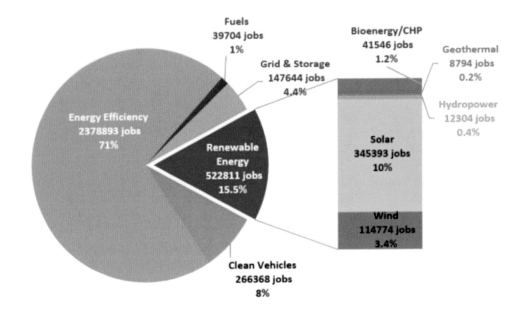

For example, the State of California accounts for about 40% of U.S. solar capacity and 31.7% of all solar jobs. Charts SE01 and SE02 illustrate the top solar energy-producing states in the U.S. Some are more aggressive than others, of course. However, what some states lack in solar energy, they make up in other locally available renewable sources such as wind power, Geothermal and Biofuel. An additional benefit of solar power is the fact that over 40% of jobs are held by women. This is a positive shift from the previously men dominated energy sectors.

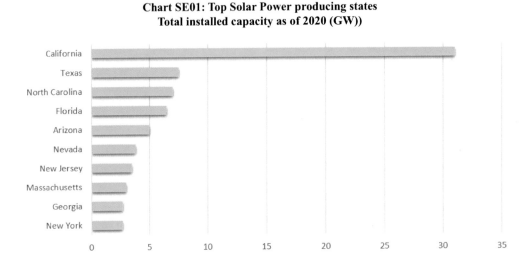

Chart SE02: The state of California produced over 40% of all U.S. solar power generated in 2020. The top ten states generated over 95% of all solar power generated in the U.S. in 2020.

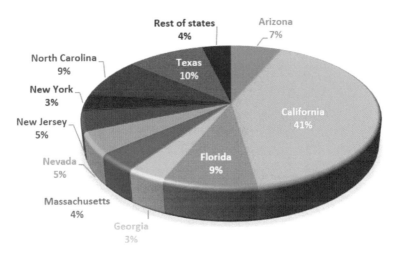

Chart SE03: Globally, China outpaces any nation with over 250 GW of total solar power capacity installed.

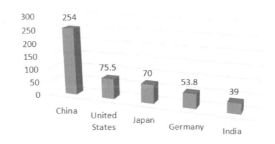

Types of solar panels

Powering small devices, such as calculators, watches and other small electronic devices with small photovoltaic (PV) cells, was generally common knowledge for a very long time. However, photovoltaic effect, which is the science of generating electricity with solar cells, was first discovered in 1839 by Edmond Becquerel. Bell Lab's development of the first silicon PV cell in 1954 elevated the technology and allowed PV cells to power larger devices over extended periods of time. Photovoltaic (PV) gets its origin from (photon), which is the process of converting light to (voltage), which is electricity. Thus, photovoltaics effect is the process of converting light into electricity. This process requires multiple hardware, of which solar cells, made from semiconduction materials, are the critical component. Numerous solar cells put together are called solar panels or solar modules. In addition to solar cells, solar panels consist of metal frames and glass casing units for surface protection. They also consist of wiring cables to transfer the electric current from the semiconducting cells to the devices to be powered. Further development led to the arrangements of multiple panels, put together in arrays to generate electricity on a larger scale.

Silicon Solar Panel

Silicon solar panels: are the most used solar panels, occupying more than 90% of the global PV market. These panels vary in their manufacturing process, appearance, performance and costs. They have an efficiency (the rate at which the solar cell converts sunlight into electricity) of up to 33%. As the name indicates, silicon solar panels are made from silicon, which is the most common semiconductor material used today. Silicon is one of the most abundant nonmetal elements. In fact, more than 90% of the Earth's crust comprises silicate minerals, making silicon the second most abundant element in the Earth's crust after oxygen.

Silicon has the second highest melting and boiling points of the non-metals. Thus, silicon based panels can withstand harsh elements, whether it be from the sun's heat or from freezing temperatures. More importantly, exposing silicon to the sun breaks its molecular bond leading to loosened protons and electronics. The loosened electrons are then captured and used as electricity. Silicon solar panels are basically silicon cells covered with a glass sheet and framed together. Putting the cells in panels makes transporting and installing them much more manageable. Solar panels are installed on roofs of residential or commercial buildings or deployed on ground-mounted racks to create massive, utility-scale systems. Solar panels come in standardized 60, 72, and 96 cell variance.

Silicon cells are made from either **monocrystalline** or **polycrystalline** silicon crystals. Monocrystalline solar cells are manufactured from a single silicon crystal. Therefore, they have the highest efficiency and power capacity. They can reach efficiencies of higher than 20 percent and have a power capacity of more than 400 watts (W) and few exceeding 500 W. Monocrystalline cells are distinguished from others by their dark color, which is the appearance of pure silicon crystals when exposed to sunlight.

Manufacturing monocrystalline cells is an energy-intensive process and results in wasted silicon material. Thus, these cells are very expensive. Fortunately, the silicon waste is processed to manufacture other cells, referred to as polycrystalline solar cells.

Polycrystalline solar cells are composed of fragments of silicon crystals melted together in a mold before being cut into wafers. Thus, they are the least efficient and less expensive. These panels usually have efficiencies of around 17 percent and up to 300 W power capacity. The panels appear bluish due to how light reflects off the cell's silicon fragments.

Thin-Film solar panels: Another commonly used photovoltaic technology is thin-film solar cells because the cells are made from thin layers of semiconductor material. Thin-film cells are made from a variety of materials, most commonly from cadmium telluride (CdTe). Manufactures make this type of thin-film panel by placing a layer of CdTe between transparent conducting layers that help capture sunlight. This type of thin-film technology also has a flexible glass layer on top for protection.

Another material used in the manufacturing of thin-film solar panels is amorphous silicon (a-Si), which is similar to the composition of monocrystalline and polycrystalline panels. Though these thin-film panels use silicon in their composition, they are not made up of solid silicon wafers. Instead, they're composed of non-crystalline silicon placed on top of glass, plastic or metal. Lastly, Copper Indium Gallium Selenide (CIGS) panels are another popular type of thin-film technology. Electrodes are placed on the front and the back of the material to capture electrical currents.

Overall, thin-film solar cells are flexible and lightweight and ideal for portable applications, such as backpacks, RVs and boats. Their lighter weight and flexibility contribute to a lessor, labor-intensive installation leading to a cheaper installation cost. These cells, however, have the least efficiency at about 11 percent and thus, are the least expensive. Thin-film technology does not come in uniform sizes, unlike monocrystalline and polycrystalline solar panels. As such, the power capacity from one thin-film panel to another will largely depend on its physical size. As far as color goes, thin-film solar panels come in blue and black hues, depending on the material used.

Bifacial solar panels: These panels capture sunlight from both the front and back of the panel, thus producing more electricity than comparably sized, traditional solar panels. Unfortunately, installation has its own challenges. Bifacial modules mounted flush on a rooftop block any reflected light from reaching the backside of the cells. Thus, the higher a bifacial module is tilted, the more power it produces. Therefore, ground-mounted installation is preferred because there is more room for tilt allowing light to reflect on both sides of the panel.

Bifacial solar panels are typically manufactured with monocrystalline solar cells, but polycrystalline bifacial solar panels exist as well. It is estimated that bifacial modules lead to a 10-30% increase in production over monofacial panels. The increased production capacity is reflected in their higher purchasing cost.

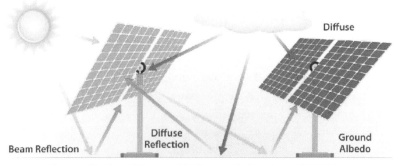

Source: National Renewable Energy Laboratory (NREL)

III-V solar panels: III-V solar cells are mainly constructed from Group III elements; gallium and indium, and Group V; arsenic and antimony of the periodic table. These solar cells are generally much more expensive to manufacture than other technologies. But they convert sunlight into electricity at a much higher efficiencies of up to 47%. Because of their high efficiency rate, these solar cells are often used on space satellites, unmanned aerial vehicles and other applications that require a high ratio of power-to-weight.

Next-Generation solar cells: There are ongoing research and development worldwide on many new photovoltaic technologies and materials, including organic materials, quantum dots and hybrid organic-inorganic materials with a promise of lower cost and higher efficiency panels. In addition, continuous improvement of what is already available contributes to solar power becoming more reliable, efficient and cheaper. According to the International Energy Agency (IEA), solar power is expected to become the world's biggest source of electricity by 2050.

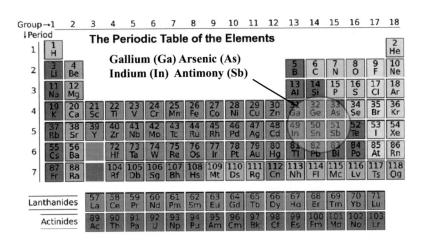

Additional hardware

In addition to solar panels, generating solar power by photovoltaic requires other hardware, including inverters, charge controllers, batteries, wirings and framing.

Inverters: An inverter is a piece of hardware connected to the solar panel by copper wiring. Its purpose is to convert the direct current current (DC) generated by the panels into usable alternating current (AC). There are three different solar power inverter technologies:

Microinverters: are attached to each panel and each panel is monitored and optimized separately. The energy produced is converted to AC at each panel rather than sent to a central location. Thus, a failure in a single panel does not impact the other panels. Microinverters are the most efficient and, as a result, the most expensive inverters. Adding to their cost is also the challenges of maintenance or repair because of the many different locations these inverters are installed.

Macroinverters/String Inverters: are used in a centralized way rather than the individualistic approach of the microinverters. In this system, each panel is wired together into a string, and multiple strings are connected to a single central inverter, usually located at ground level. The central inverter will convert all of the electricity from the solar panels into AC. Thus, a failure in one panel can impact the performance of the overall array of panels. The output is as efficient as the least productive panel. Macroinverters are preferred by most small-scale solar power projects due to their affordability. Maintainance and repair are also less challenging since there is only one central location, mostly at ground level.

Power Optimizer Inverters: combine both technologies of micro and macro inverters. They are considered as efficient as microinverters but at a slightly cheaper cost. Like microinverters, power optimizers are located at each solar panel. This concept helps improve the efficiency of the panels. However, like macroinverters, the energy is sent and converted to AC at a centralized location, reducing maintenance and repair costs.

In addition, inverters are responsible for monitoring and optimizing the system. Thus, they serve the purpose of an online communication portal. They provide system information on the amount of solar energy being produced. They also offer an inside look into the system to ensure it is functioning correctly.

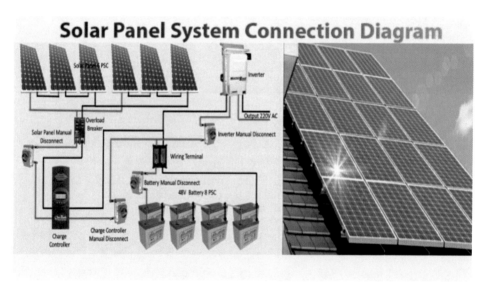

Charge controllers: Charge controllers are a necessity for a battery-backed solar power system where excess power is stored in a battery for later use. They act like an on and off switch, allowing power to pass when the battery needs charging and cutting it off when the battery is fully charged. They also contribute to the life span of the battery by improving charge quality. They prevent the battery from being overcharged. They also prevent the battery from discharging in the absence of sunlight.

Batteries: Solar panels can only produce power when the sun is shining. Even then, the amount of sunlight varies depending on location, time of day, the season of the year, and weather conditions. Thus, storing unused daytime energy for later use is increasingly becoming important.

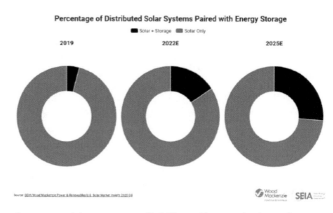

Batteries provide that storage thus, provide energy reliability. Currently, less than 5% of solar power systems are paired with battery storage. It is estimated that the number will grow to more than 25% by 2025 as more homeowners and businesses are increasingly demanding solar systems that are paired with battery storage. Batteries can be arranged in many different arrays to provide storage for all scales of solar panel installations.

Solar power installations

Solar PVs are installed in various scales including, residential, commercial, community and utility. At the utility level, however, solar generates the cheapest electricity of any renewable or fossil fuel-based power.

Residential-scale solar power: typically installed on rooftops of homes or in open land (ground-mounted) and are generally between 5 and 20 kilowatts (KW), depending on the property's size. The average U.S. residential solar installation is about 5 KW or around 20 panels. Unfortunately, the average cost of solar power installation in an average U.S. residence is $20,000. By the time financing or leasing costs are added, it can take decades to recoup the benefits.

Commercial-scale solar power: generally installed at a greater scale than residential solar. Though individual installations can vary significantly in size, commercial-scale solar power provides on-site solar power to businesses. Data from SEIA's annual "Solar Means Business" report show that major U.S. corporations, including Apple, Amazon, Target and Walmart, are investing in solar energy at an incredible rate. Through 2018, the top corporate solar power users in the U.S. have installed more than 7 GW of capacity across the country in more than 35,000 different facilities. More than half of the 7 GW of corporate solar capacity has been installed in the last three years.

Net-metering: Both residential and commercial solar power projects can benefit from Net-metering. In simple terms, Net-metering is a solar incentive program that allows consumers to store excess solar power in the utility electric grid in exchange for credits. Then, at night or other times when the solar panels are under producing, consumers pull energy from the grid and use the credits to offset energy costs. The program saves utility companies from having to generate additional power during peak times while rewarding consumers for their unused solar power. This is called a "Grid-tie" or "On-grid" solar power system. For comparison, "Off-grid" system is where residents chose to set up a stand-alone solar power system, keeping their homes off the utility company's grid and generating power solely for their own use.

Community-scale solar power: is a viable solar option for solar enthusiasts who can not or choose not to install solar panels on their property. Community-scale solar projects are typically built in a central location allowing residential consumers to subscribe and receive many of the benefits of solar power without installing solar panels on their property.

Utility-scale solar power: These projects are typically large installations that provide solar power to a large number of utility customers. Most utility-scale solar projects are installed on land, with few exceptions where they are installed on water or sea surface. These installations are knowns as floating solar farms. They are a great alternative in areas where land space is a challenge, such as in Singapore. In addition to a cheaper and abundant source of energy, what does solar power mean for the environment? According to the Environmental Protection Agency (EPA)'s Greenhouse Gas Equivalencies Calculator, the average American home going solar for a year is like:

- Reducing carbon dioxide emission by more than 12,500 pounds.
- Not burning over 8,000 pounds of coal.
- Driving about 18,000 miles less.
- Not charging 937,683 smartphones.

Solar thermal (heat) energy

The second way of utilizing active solar energy is solar thermal/heat energy. It is a process of using panels to capture the sun's heat (rather than light) and use it directly for heating fluids or space. It can be utilized on a smaller scale to heat residential and business spaces or on larger utility-scale projects to generate electricity for many.

Small scale solar thermal energy: utilizes many different technologies to collect and convert solar radiation into usable heat energy for various purposes, including for heating the interior of buildings, greenhouses and even swimming pools. For example, solar water heating systems are composed of solar collectors, insulated piping and hot water storage tanks to collect the sun's thermal energy. Then, the accumulated energy or hot water is circulated with either an electric pump (active) or using gravity (passive). Active solar water heating systems are more common in residential and commercial use. Passive solar water heating systems are typically less expensive, but they are also less efficient.

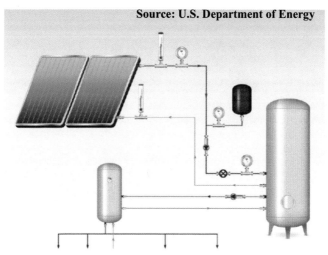

Source: U.S. Department of Energy

Passive water heating system: The heated water is distributed using gravity. No electric pumps required

Larger or utility-scale solar thermal energy: This process uses many panels to collect heat to be used to boil water or other fluids. The steam from the boiling fluid is used to spin a large turbine/generator to produce electricity. This method is called **concentrated solar power (CSP) system**. The process harnesses heat from the sun to provide electricity for large power stations. For example, Solar Star, America's largest solar farm, produces 579MW using 1.7 million solar panels over 3,200 acreages of land space.

There are three main types of CSP systems: linear concentrator, dish/engine and power tower systems.

Linear concentrator systems collect the sun's energy using long rectangular, curved (U-shaped) mirrors. The mirrors are tilted toward the sun to expose attached tubes or receivers to maximum sunlight. The reflected sunlight heats a fluid flowing through the tubes. The hot fluid is then used to boil water in a conventional steam turbine generator to produce electricity.

Linear concentrating collector fields consist of a large number of collectors in parallel rows that are typically aligned in a north-south orientation to maximize annual and summer energy collection. In these systems, the collector field is oversized to heat a storage system during the day, so the additional steam it generates can be used to produce electricity in the evening or during cloudy weather.

There are two major types of linear concentrator systems: parabolic trough systems, where receiver tubes are positioned along the focal line of each parabolic mirror; and linear Fresnel reflector systems, where one receiver tube is positioned above several mirrors to allow the mirrors greater mobility in tracking the sun.

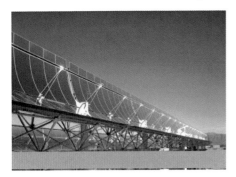
Parabolic trough systems

Linear Fresnel reflector systems

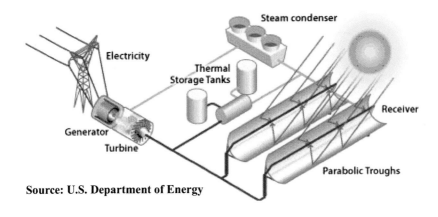
Source: U.S. Department of Energy

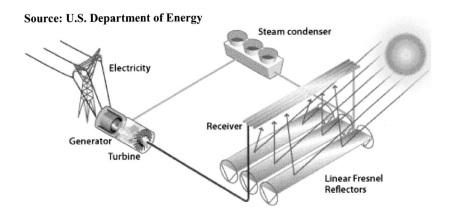

Source: U.S. Department of Energy

Dish/Engine systems use a mirrored dish similar to a large satellite dish. The mirrored dish is usually composed of many smaller flat mirrors formed into a dish shape to minimize cost. The dish-shaped surface directs and concentrates sunlight onto a thermal receiver, which absorbs and collects the heat and transfers it to the engine generator. The most common type of heat engine used in dish/engine systems is the Stirling engine. This system uses the fluid heated by the receiver to move pistons and create mechanical power. The mechanical power is then used to run a generator or alternator to produce electricity.

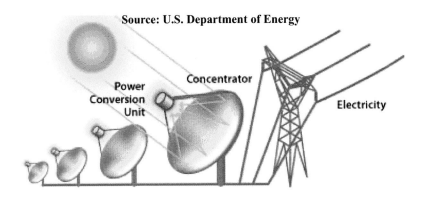

Source: U.S. Department of Energy

Power Tower systems: A power tower system uses a large field of flat, sun-tracking mirrors known as heliostats to focus and concentrate sunlight onto a receiver on the top of a tower. A heat-transfer fluid heated in the receiver is used to generate steam, which, in turn, is used in a conventional turbine generator to produce electricity. Some power towers use water/steam as the heat-transfer fluid. Other advanced designs are experimenting with molten nitrate salt because of its superior heat transfer and energy storage capabilities. The energy storage ability or thermal storage, allows the system to continue to dispatch electricity during cloudy weather or at night.

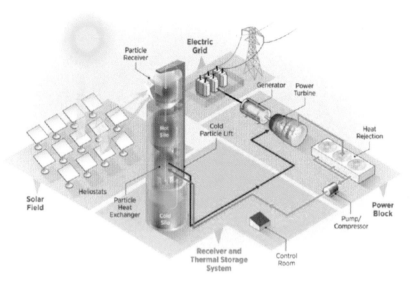

The Ivanpah Solar Electric Generating System is the largest concentrated solar thermal plant in the world. Located in California's Mojave desert, the plant is capable of producing 392 MW of electricity using 173,500 software controlled mirros, called heliostats. Each heliostats has two mirrors that follow the sun and reflect it onto water-filled boilers atop three separate 450 foot towers. Construction began in 2010 & completed 2014 at a cost of $2.2 billion USD. The site covers an area 5.5 mi^2. When the sunlight hits the boilers, the water inside is heated and creates high temperature steam. The steam is then piped to conventional steam turbines, which generate electricity.

Aside from the U.S., Spain has several power tower systems. Planta Solar 10 and Planta Solar 20 are water/steam systems with capacities of 11 MW and 20 MW, respectively. Gemasolar, previously known as Solar Tres, produces nearly 20 MW of electricity and uses molten-salt thermal storage.

Geothermal Energy

The word "Geothermal" comes from the Greek words geo, meaning earth and thermal, meaning heat. Thus, geothermal means heat energy that exists within the earth. It exists in reservoirs of hot water and steam deep within the sub-surface of the earth. In simpler terms, thermal energy originates from the original formation of the planet and the slow decay of radioactive particles in the earth's core. It is believed that the earth's internal temperature at the core can reach up to 10,000 ^{0}F (5,537 ^{0}C), which is as hot as the sun's surface. These extreme temperatures, coupled with the high pressure at the core, melt any solid material, such as iron and rocks, into a glue-like viscous fluid. Since the resulting fluid is lighter, it start to rise upwards towards the earth's crust in the form of volcanoes and lava. This is the source of geothermal energy that we have exploited for generations.

Archaeological evidence shows geothermal use by all ancient cultures around the world. For example, there are evidence of Native Americans using hot springs for heating, cleaning and healing as far back as 10,000 years ago.

The recorded history of geothermal energy use, however, goes back 2000 years. The Chinese, Romans and Turkish are among the many cultures where hot springs were used for bathing and heating. In modern times, geothermal energy was used for district heating in the late 1400s in France followed by similar district heating system in the America's in Boise, ID in 1890s.

The early 1900s saw the rise of geothermal energy for the purpose of heating buildings andgreenhouses. Hot Lake Hotel, in Union County, Oregon, became the first known building in the world to use geothermal energy to heat the building in 1907. This was followed by the consumption of geothermal heating of greenhouses in Boise, Iceland and Italy in the 1920s. The early 1940s witnessed the use of steam and hot water from geysers for heating residential homes in Iceland. Today, Iceland is the world leader in this practice, with over 90% of its homes heated with geothermal energy, saving the country over $100 million annually in oil imports. Reykjavík, Iceland has the world's biggest district heating system, often used to heat pathways and roads to hinder ice accumulation.

The early 1900s also witnessed the rise of geothermal energy for electricity production. The first known geothermal power plant was tested in Italy in 1911, followed by the first industrial geothermal electricity production in 1958 in New Zealand. Today technological advance is allowing many nations to utilize geothermal energy to generate electricity worldwide.

Source of geothermal energy

To understand why geothermal energy is considered renewable, we need a basic understanding of the earth's interior formation. The earth has four major layers of different materials, such as iron, nickel and silicate. The extreme temperature and pressure that occurs in the earth's interior affects the melting point of these materials and result in the earth's layers' different characteristics.

- **An inner core** is the deepest layer made up of solid iron/nickel that is about 1,500 miles (2414km) thick. The temperature of the earth's inner core ranges from 7,000 °F to 11,000 °F, which is as hot as the surface of the sun. The intense pressure that exists in this layer changes the melting point of the nickel-iron, allowing it to remain solid, despite the extreme temperature.

- **An outer core** is an outer layer of the core made up of hot molten iron/nickel called magma. It is about 1,500 miles (2414km) thick. This layer is liquid due to the high temperatures and the decreased pressure compared to the inner core.

- A **mantle** is a layer of magma and rock surrounding the outer core. It is about 1,800 miles (2890km) thick making it the thickest layer. Temperatures in the mantle range from about 392°F at the upper

- mantle-crust boundary to approximately 7,230°F at the mantle-outer core boundary. The mantle layer contains solid/glue-like silicate material. The upper mantle is fairly solid silicate due to lower temperature and pressure, while the lower mantle is molten and has lower viscosity due to increased temperature. The heat from this layer escapes and surfaces up towards the crust through fractured or faulted tectonic plates heating water in the pores and fractures of the rocks. A small portion of the heated water/steam rises to the surface creating hot geysers and hot springs.

- A **crust** is the outermost layer made up of solid rock that forms the continents and ocean floors. The layer is 15 to 35 miles (24km-56km) thick under the continents and 3 to 5 miles (5km-8km) thick under the oceans. As previously mentioned, silicate materials make up over 90% of the earth's crust, while nickel/iron alloys make up the rest. The low temperature and pressure that exists at this layer allow the crust to be solid.

The earth's crust is broken into pieces called tectonic plates. These plates under the continents and ocean floors drift apart and push against each other at the rate of about one inch per year in a process called continental drift. Magma comes close to the earth's surface near the edges of these plates, where the crust is faulted or fractured by the continuous drift and exits in the form of a volcanic eruption. The lava that erupts from volcanoes is partly magma.

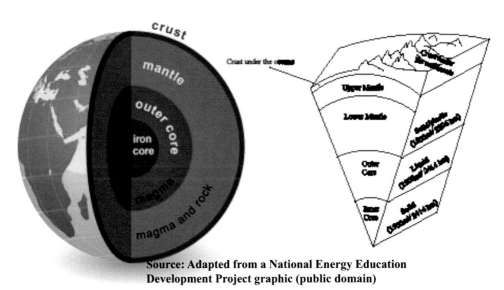

Source: Adapted from a National Energy Education Development Project graphic (public domain)

Since magma flows upwards to near the edge of the tectonic plates, it makes sense to look for geothermal energy along these lines. Scientists have identified seven major tectonic plates along with many more minor and micro plates. The seven major plates are North American, South American, Pacific, African, Antarctic, Eurasian and Indo-Australian. Most large scale geologic events, such as volcanoes or earthquakes, occur along these tectonic plates along the edges of the continents, island chains and beneath the sea. In fact, more than half of the world's active volcanoes above sea level encircle the Pacific Ocean to form the circum-Pacific "Ring of Fire".

Regarding geothermal energy, most of the previously discovered and/or the untapped thermal energy exists along these lines. For example, two of the major moving tectonic plates (the North American and the Pacific plates) meet on the coast of California. The boundary where they meet is a major earthquake area, known as the San Andreas fault. The entire San Andreas fault system extends over 800 miles long and at least 10 miles deep inside the earth's interior making California, the top geothermal energy producer in the world.

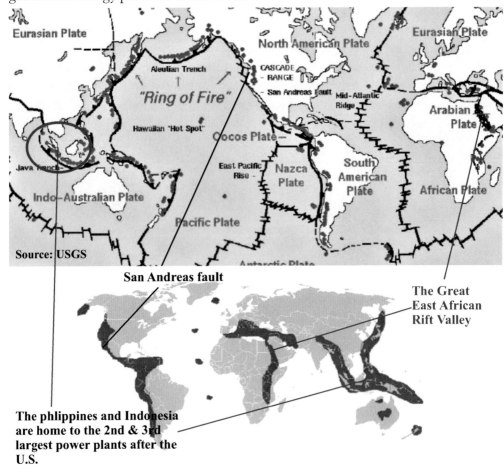

San Andreas fault

The Great East African Rift Valley

The phlippines and Indonesia are home to the 2nd & 3rd largest power plants after the U.S.

Geothermal Energy

Another area where there is a huge untapped geothermal energy source is in the East African Rift System (EARS), where the African plate and the Arabian plate meet. The rift extends thousands of miles from Lebanon in Asia, past the Red Sea before turning inland into the Danakil region of Eritrea and Ethiopia., making the region one of the most active volcanic areas on earth. The rift continues towards the Ethiopian highlands into Kenya, Uganda, the southern tip of South Sudan, and Rwanda. Finally, the rift runs along two separate branches (Western and Eastern Rift Valley), touching Tanzania, Zambia and Mozambique. The EARS has allowed Kenya and Ethiopia to take advantage of the available geothermal resource. In fact, Kenya generates 46% of its electricity from geothermal energy. Other nations, such as Eritrea and Uganda, are heavily involved in research and development to tap into this abundant energy source.

Types of extraction

Since heat is continuously produced inside the earth, geothermal energy is a great source of renewable energy. As previously mentioned, geothermal energy has been traditionally used for heating. However, harnessing geothermal energy to generate electricity is a relatively new technology and remains untapped potential that can contribute to a renewable energy mix portfolio.

Heating and cooling

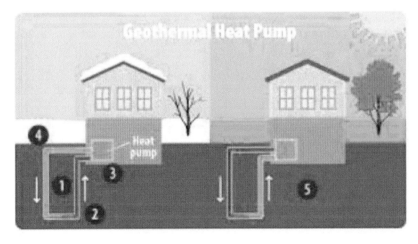

Thermal energy from underground reservoirs where heat temperatures are between 80-300 °F (26–148 °C), is directly used for heating in homes, industries, greenhouses and for bathing in many places. In addition, the thermal energy from shallow reservoirs, where temperatures are between 50-70 °F (10 to 20 °C), is extracted with heat pumps for similar purposes of space heating. The process requires digging wells directly into natural hot springs, geysers or underground where the temperature fluctuation is minimal.

Because of its long history of application, geothermal heating is more cost-effective than geothermal power (electricity generation). As a result, geothermal heating has increasingly become the preferred energy choice for home heating. Typically, more than half of geothermal energy consumption is for space heating, while another third is used for heating pools. The remainder is used to support industrial and agricultural applications.

Electricity

Technologies for direct heat consumption like space heating, greenhouses and other applications are widely used and considered mature. But when it comes to generating electricity, the temperature of the thermal reservoir is the deciding factor. There are three geothermal power plant technologies used today.

Dry steam power plant: This technology involves using steam directly from high temperature reservoirs to help spin a turbine that is attached to an electricity-producing generator. This technology has been used since 1904. The largest geothermal power facility, The Geysers geothermal plant complex in California is a dry steam facility.

Flash steam power plant: This technology uses extracted hot water (greater than 360°F/182°C) and is pumped under high pressure into a tank at the surface. The tank is held at a much lower pressure, causing some of the fluid to rapidly vaporize or "flash." The vapor then drives a turbine, which drives a generator. If any liquid remains in the tank, it can be flashed again in a second tank to extract even more energy. Flash steam plants are the most common type of geothermal power generation plants in operation today.

Binary cycle power plant: This technology allows for the generation of electricity from lower temperature reservoirs profitably. Binary cycle power plants were developed in the Soviet Union in the 1960s and allowed for electricity to be generated from much lower temperature resources than the previously used heat pumps of the 1940s. In a binary cycle power plant, water is actively injected into wells to be heated and pumped back out. The heated water passes through a heat exchanger in a Rankine cycle binary plant and vaporizes the working fluid that drives a turbine. Because resources below 300°F represent the most common geothermal resource, a significant proportion of geothermal electricity in the future could come from binary-cycle plants. Environmentally, Binary cycle power plants are closed-loop systems, thus, virtually nothing (except water vapor) is emitted to the atmosphere.

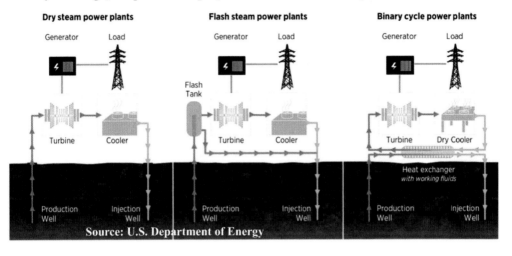

Source: U.S. Department of Energy

Enhanced Geothermal Systems (EGS):

Improvements in drilling and extraction technology allows electricity production over a much greater geographical range, including in extensional terrains, where heating occurs via deep circulation along faults, such as in the Western United States. Water is injected under high pressure to expand existing rock fissures to enable it to flow in and out freely. The technique, which is adapted from the oil and gas industry, allows for wells as deep as 6 miles (10km) compared to previous depth of 2 miles (3km). This new technique is called **enhanced geothermal systems (EGS)** and is commonly used worldwide today.

According to the 2016 Geothermal Energy Association (GEA) report, the United States was the largest geothermal electricity producer at 3.4 GW from over 70 power plants, including the largest group of geothermal power plants in the world located at The Geysers geothermal field in California. As shown on the below map, there are only few economically favorable places in the U.S. where electricity can be generated by geothermal energy. Those places are mostly located along the San Andreas fault in the western part of the country.

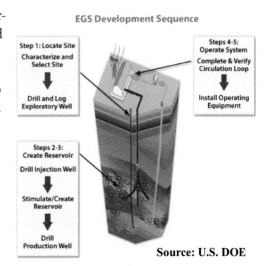

Source: U.S. DOE

Although there are few additional regions that can utilize geothermal energy for electricity production, there are only seven states that account for 100% of the U.S. production. Obviously, the tectonic nature of these U.S. teritories gives them an advantage over the other states.

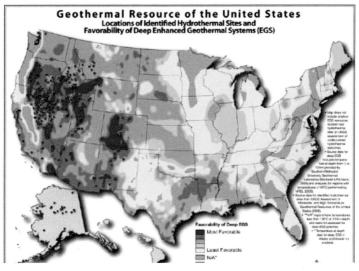

Electricity generation requires high or medium temperature resources which are located close to tectonically active regions marked by red on this map.

Considering geothermal fluids do not reach the high temperatures of steam from boilers, thermal efficiency of geothermal electric plants is low, around 10–23%. In

	State share of total U.S. geothermal electricity generation	Geothermal share of total state electricity generation
California	70.5%	6.1%
Nevada	24.5%	10.2%
Utah	2.1%	1.0%
Hawaii	1.2%	2.2%
Oregon	0.9%	0.2%
Idaho	0.5%	0.5%
New Mexico	0.3%	0.2%

addition, plant capital cost, such as drilling and exploration of deep resources account for over half the costs. Operational cost to run specialized pumps add more risk to the project. Thus, geothermal energy is currently for only those who are blessed with unstable tectonic plates beneath their feet.

The Geysers Geothermal Complex: Located in California, USA, these fields produce the largest geothermal power plant in the world. The facility draws steam from more than 350 wells spread over a 45 square miles territory. At its peak, the facility had a capacity of 1.2GW and 22 power plants, each producing 20 to 120MW. The facility generates enough electricity to power nearly one million homes.

Unlike most geothermal resources, the Geysers is a dry steam field that mainly produces superheated steam. Gravity and seismic studies suggest that the source of the heat for the steam reservoir is a large magma chamber over 4 miles (6.4 km) beneath the surface and greater than 8 miles (13 km) in diameter.

Historical records show that Native Americans were the first to utilize these Geysers for heating, bathing and cooking for over 12,000 years. Later, its steam baths operated as a resort in the 1800s. The arrival and westward expansion of Europeans and European Americans in the mid-1800 further developed the area commercially. During the 1850s, "The Geysers Resort Hotel", a hot spring spa, was opened. In 1960, Pacific Gas and Electric began operation of their 11-megawatt geothermal electric plant at the Geysers. Today, the Geysers are the largest geothermal power plant in the world producing about 9% of California's renewable energy mix.

Globally, the International Geothermal Association (IGA) has reported that 10.7 GW of geothermal power in 24 countries is online, which was expected to generate 67,246 GWh of electricity in 2010. This represents a 20% increase in online capacity since 2005. IGA projects growth to 18.5 GW by 2015, due to the projects presently under consideration, often in areas previously assumed to have little exploitable resources. Worldwide, 11.7 GW of geothermal power was available in 2013, spread out globally from Iceland to New Zealand, El Salvador, and Kenya.

Geothermal power plant

After the United States, the Philippines is the second-highest producer of geothermal energy, with 1.9 GW of capacity making up approximately 27% of Philippine electricity generation. Indonesia has the third-largest production at 1.6 GW. Indonesia also has the most extensive geothermal reserve in the world, capable of producing 20GW. Indonesia is predicted to be the largest geothermal electricity producer within a decade. Additionally, there are many volcanic areas of the world, including the "Great Rift Valley" in east Africa, which posses an excellent energy potential under the right technology.

Renewability and sustainability

Geothermal energy is considered renewable because heat is continuously being produced below the earth's surface. It is sustainable because it returns an equal volume of water to the area that the heat extraction takes place. For example, high temperate water/steam is extracted from underground reservoir and cooler water is replenished back to the underground reservoir. The thermal/heat difference between the .waters is the energy gained. The extraction must still be monitored to avoid local depletion.

Geothermal energy worries

Depletion: The three oldest sites, at Larderello, Wairakei and the Geysers, have experienced reduced output because of local depletion. Depletions occur when steam and water is extracted faster than they are replenished. For example, the Geysers geothermal complex in California were producing 1.5GW of electricity with 22 power units during their peak. Today, they are reduced to 900MW with 18 units because of depletion. Fortunately, reducing extraction and reinjecting more water can help the wells recover their full potential. These recovery strategies have been implemented successfully at some sites, including the Lardarello field in Italy, at the Wairakei field in New Zealand and at the Geysers in California. All three have been operational since 1913, 1958, and 1960 respectively.

Environmental Concerns: Although the low emissions of geothermal energy are considered to have excellent potential for the mitigation of global warming, there are some environmental effects to keep in mind. They include:

- Geothermal wells release greenhouse gases trapped deep within the Earth, although the emissions are much lower than fossil fuel emissions.

- Fluids drawn from the deep Earth carry a mixture of gases, notably carbon dioxide (CO_2), hydrogen sulfide (H_2S), methane (CH_4) and ammonia (NH_3). These pollutants contribute to global warming, acid rain and noxious smells if released. Plants that experience high levels of acids and volatile chemicals are usually equipped with emission-control systems to reduce the exhaust.

- Hot water from geothermal sources may also contain trace amounts of toxic elements such as mercury, arsenic, boron and antimony. These chemicals precipitate as the water cools and can cause environmental damage if released.
- The required hardware, such as pumps and compressors, may consume energy from a polluting source but is a fraction of the heat output.
- Enhanced geothermal systems can trigger earthquakes as part of hydraulic fracturing. For example, a project in Basel, Switzerland, was suspended because more than 10,000 seismic events measuring up to 3.4 on the Richter scale occurred over the first six days of water injection.
- Drilling holes and flooding with water may have unintended consequences of bedrock destabilization that leads to earthquakes, similar to what is witnessed in fracking.

Benefits of geothermal energy

- **Renewable:** Through proper reservoir management, the rate of energy extraction can be balanced with a reservoir's natural heat recharge rate.
- **Baseload:** Geothermal power plants produce electricity consistently regardless of weather conditions.
- **Small Footprint:** Geothermal power plants are compact, using less land per GWh (404 m^2) than coal (3642 m^2), wind (1335 m^2) or solar PV (3237 m^2). Geothermal also has minimal land and freshwater requirements (5.3 US gal) than other plants like nuclear, coal or oil (260 US gal) per MWh.
- **Clean:** Modern closed-loop geothermal power plants emit no greenhouse gasses; life cycle GHG emissions are four times less than solar PV and six to twenty times lower than natural gas.
- **Scalability:** Geothermal power is highly scalable, operating from a rural village to an entire city. The power generation is local and requires no lengthy and expensive grid system.

As shown above, geothermal energy requires careful planning to be a sustainable and environmentally a safe energy source. Also, since it is only profitably available in few places on earth, it may never be a significant energy source for many. Seven countries are responsible for nearly 80% of global geothermal power plant capacity. However, geothermal energy can be part of a renewable energy portfolio mix. Considering geothermal energy is available under any weather conditions day and night, it can also be used as energy storage to supplement other sources such as wind and solar.

The Danakil Depression, Eritrea
(untapped geothermal resource)

The small nation of Eritrea is rich in culture and history. Located on the coast of the Red Sea in the horn of Africa, it sits in an area where humans are believed to have migrated from Africa to Asia & the rest of the world. It also has a rich history of being part of the old African kingdoms of Punt, Kush and Axum. Politically, however, this small nation is relatively new on the world stage, having gained its independence in 1991.

Following independence, the government of Eritrea partnered with many experts to develop its energy sector, which is key to any economic advancement. After all, the country is blessed with so many natural energy resources. For example, Eritrea's 1100km of Red Sea coastline has proven petroleum and natural gas resources. Its arid & hot lowlands are great potential for solar energy. The highlands offer a great opportunity for wind energy. Finally, the Danakil region are great source of geothermal energy. Unfortunately, these energy potentials have not been realized, yet. Geopolitical instability has taken its toll on this young nation.

In 1996, the Eritrean government partnered with the United States Agency for International Development (USAID) and the United States Geological Survey (USGS) to exploit its geothermal resources in the Danakil depression region.

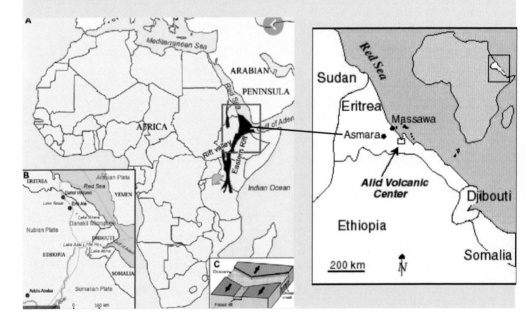

The Danakil Depression is a vast plain, 200 km by 50 km (124 by 31 mi), lying in the north of the Afar Region of Eritrea and Ethiopia. It is about 125 m (410 ft) below sea level, making it the shallowest place on earth. Geologically, the Danakil Depression lies at the triple junction of three tectonic plates; the African, Arabian and Indian tectonic plates. Geologists explain its origins as a result of Africa and Asia moving apart, causing rifting and volcanic activity.

The study was more focused on the Alid volcanic center. Extensive studies had taken place to produce a geothermal resource assessment based on mapping of the geology and study of the fumarolic emissions from the Alid volcanic center.

The study concluded that "compositions of fumarolic gases collected at Alid indicate that the reservoir temperature of a hydrothermal-convection system driven by this heat source is very likely in the range of 250° to 300°C. The overall temperature and permeability conditions seem so favorable for an electrical grade geothermal resource that exploration drilling to depths of 1.5 to 2 km is recommended". (source: https://pubs.usgs.gov/of/1997/0291/report.pdf)

Unfortunately, geopolitics took a toll on this young nation. Any project along with economic advancement halted in 1998, following a war with neighboring Ethiopia and economic sanction by the U.S. and European Union. Eritrea was again forced to forfeit its economic potential in order to defend itself and its existence on the world stage. Fortunately, recent trends show a promise of stability which will allow the country and its citizens to shift their attention towards economic development once more. As previously mentioned, Eritrea is blessed with all kinds of renewable and non-renewable energy sources. The only natural resource it lacked is water. The government has managed to build over 700 water dams and reservoirs of all sizes in the past thirty years. These reservoirs can someday serve as pumped storage and divergence types of hydro energy and add to a vast renewable energy mix for this young nation.

Bioenergy

Bioenergy is renewable energy derived from biological sources. It is a form of renewable energy that is derived from recently living organic materials known as biomass. Biomass is any organic material that has stored sunlight in the form of chemical energy. In simpler terms, Biomass is the fuel, and bioenergy is the energy extracted from the biomass/fuel. Biomass includes wood, wood waste, crops, crop residue like straw, food waste, animal byproducts/manure, sugarcane, and much more agricultural products and agricultural residues.

Biomass fuel was initially produced as a byproduct, residue, or waste-product of other industrial processes, such as farming, forestry, and food processing. Recently though, biomass fuel has been agriculturally grown specifically for biofuel production. These agricultural products include corn and soybeans in the United States; rapeseed, wheat, sugar beet, and willow in Europe; sugarcane in Brazil; palm oil in Southeast Asia; sorghum and cassava in China; and jatropha in India. Additionally, other biomass fuel sources are being developed. These include municipal and household waste, also known as Sewage biomass.
Biomass is mainly used in three different forms:

Biopower: fuel for power plants to generate electricity
Biofuel: fuel for the transportation industry
Bioproducts: starter ingredient for the manufacturing industry

BIOPOWER for power utilities

Biopower technologies convert renewable biomass fuels into heat and electricity using steam and turbine processes, similar to fossil fuel-based power plants. The biomass used for electricity production ranges by region. For example, forest byproducts, such as wood residues, are popular in the United States, while rice husks are common in Southeast Asia. While Animal husbandry residues, such as poultry litter, are widespread in the UK while Brazil prefers a much abundant source of sugar cane. There are three ways to harvest the energy stored in biomass to produce biopower: burning, bacterial decay, and conversion to a gas or liquid fuel.

Burning: Most biopower is produced by burning biomass to produce high pressured steam. The steam is used to turn a turbine attached to an electricity-generating generator in a similar fashion to fossil-fueled power plants. For example, sugar and ethanol-producing plants burn the sugarcane byproduct/waste, bagasse, to provide heat for distillation. This allows the plants to be energy self-sufficient and offset the need for carbon fuels, and lowering the carbon intensity of electricity generation along the way. Excess bagasse not used as fuel is used to generate electricity, which provides additional income to the plants and additional power for utilities.

Bacterial Decay: This technology uses waste material, such as human sewage or animal dung decomposed by bacteria to produce methane gas. The gas is used to replace natural gas in generating electricity.

Conversion to gas or liquid fuel: This system exposes solid biomass to high temperatures to produce synthesis gas which is a mixture of carbon monoxide (CO) and hydrogen (H). The gas is then used in a boiler to produce electricity. Another method is to heat the solid biomass to lower temperatures under a complete absence of oxygen to produce crude-like fuel. The crude-like fuel is then used to substitute for other fuels in the turbines.

BIOFUELS for the transportation industry

As mentioned earlier, the purpose of this book is how to utilize renewable energy for electricity production. Thus, discussion of renewable energy for the transportation industry is minimal. However, brief discussions of bioenergy for transportation are necessary, considering biofuels' success in the last few decades.

Liquified Natural Gas to power commercial vehicles

Biomass that is converted into liquid fuels or gaseous fuels is known as biofuels and is used for transportation. The two most common types of biofuels today are ethanol (Bioethanol) and diesel (biodiesel). They are mainly used in airplanes, light and heavy vehicles, and ships. Biofuels offer renewable transportation fuels that are functionally equivalent to petroleum fuels and help lower the carbon intensity. Biofuels can be produced from plants, crops, commercial waste, and domestic waste. If the biomass used in the production of biofuel can regrow quickly, the fuel is generally considered to be a form of renewable energy. However, whether it is environmentally friendly or not depends on many factors, which will be discussed later in the chapter.

Bioethanol: is the most common biofuel worldwide, particularly in the United States and in Brazil. It is an alcohol made by fermentation, mostly from carbohydrates produced in sugar or starch crops such as corn, sugarcane, or sweet sorghum. The process involves enzyme digestion to release sugars from stored starches, fermentation of the sugars, distillation, and drying. The distillation process requires significant energy input for heat. The process is considered renewable only if the energy input is derived from easily replenishable sources, such as trees, grasses, bagasse, and wood chips, rather than fossil fuels.

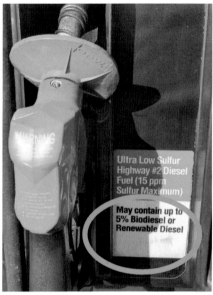

Ethanol can be used as a fuel for vehicles in its pure form (E100). However, since ethanol has a smaller energy density than gasoline, it takes much more ethanol to produce the same amount of work as gasoline. Thus, ethanol is usually used as a gasoline additive to increase octane and improve vehicle emissions. After all, ethanol does have a higher octane rating than gasoline. In 2019, worldwide biofuel production reached 161 billion liters (43 billion gallons US), contributing to 3% of the world's fuels for road transport. For those who dare to venture ahead and use biofuel for their vehicles, there is what is known as Drop-in biofuels. Drop-in biofuels are functionally equivalent to petroleum fuels and fully compatible with the existing petroleum infrastructure. They require no engine modification of the vehicle.

Biodiesel: is produced from oils or fats using transesterification and is the most common biofuel in Europe. Chemically, it consists mostly of fatty acid methyl (or ethyl) esters (FAMEs). Feedstocks for biodiesel include animal fats, vegetable oils, soy, rapeseed, sunflower, palm oil, and algae, to name a few. It can be used as a fuel for vehicles in its pure form (B100) but is usually used as a diesel additive to reduce particulates, carbon monoxide, and hydrocarbons from diesel-powered vehicles to improve efficiency. In many European countries, a 5% biodiesel blend is widely used and is available at thousands of gas stations. In the US, more than 80% of commercial trucks and city buses run on diesel. Biodiesel is also safe to handle and transport because it is non-toxic and biodegradable.

Green diesel: Green diesel is produced through hydrocracking biological oil feedstocks, such as vegetable oils and animal fats. Hydrocracking is a refinery method that uses elevated temperatures and pressure in the presence of a catalyst to break down larger molecules, such as those found in vegetable oils, into shorter hydrocarbon chains used in diesel engines. Unlike biodiesel, green diesel has precisely the same chemical properties as petroleum-based diesel. It does not require new engines, pipelines, or infrastructure to distribute and use. Unfortunately, it has not been produced at a cost that is competitive with petroleum. Green diesel is being developed in Louisiana and Singapore by ConocoPhillips, and Valero among other companies.

Straight vegetable oil: Straight unmodified edible vegetable oil is generally not used as fuel, but this truck is one of 15 based at Walmart's Buckeye, Arizona distribution center that was converted to run on a biofuel made from reclaimed cooking grease produced during food preparation at Walmart stores.

BIOPRODUCTS for the manufacturing industry

In addition to being used for biopower and biofuel, biomass can also serve as a renewable alternative to fossil fuels in manufacturing bioproducts such as plastics, lubricants, industrial chemicals, and many other products currently derived from petroleum.

Investment and jobs

The economic contribution of bioenergy, particularly in the generation of electricity, is still minimal. However, bioenergy can be lucrative in an isolated situation where waste byproducts are recycled for fuel. Deforestation and agricultural biomass competing for arable land, however, have unforeseen, long-term consequences to the overall environment. U.S. Department of Energy's 2016 Billion-Ton Report: Advancing Domestic Resources for a Thriving Bioeconomy concluded that the United States has the potential to produce 1 billion dry tons of non-food biomass resources annually by 2040 and still meet demands for food, feed, and fiber.

One billion tons of biomass could:
- Produce up to 50 billion gallons of biofuels
- Yield 50 billion pounds of bio-based chemicals and bioproducts
- Generate 85 billion kilowatt-hours of electricity to power 7 million households
- Contribute 1.1 million jobs to the U.S. economy
- Keep $260 billion in the United States.
- By 2010, there was 35 GW (47,000,000 hp) of globally installed bioenergy capacity for electricity generation, of which 7 GW (9,400,000 hp) was in the United States.

Bioenergy and the environment

The burning of carbon-based fuels always leads to carbon-based emissions regardless of whether the carbon originated from fossil fuels or biofuels. Unfortunately, many biofuel projects are not carbon neutral. Some even have higher carbon emissions than fossil-based projects. In general, any fuel or energy is considered pollutant when released into the environment at a rate faster than the environment can disperse, dilute, decompose, recycle, or store it in some harmless form. Based on this definition, both fossil fuels and some biofuels are pollutants. Thus, choosing the proper biomass is critical to adding bioenergy to our renewables energy portfolio.

For instance, in 2018, the European parliament voted to phase out palm oil use in transport fuels by 2030. A 2015 study funded by the European Commission found that palm oil and soybean oil had the highest indirect greenhouse gas emissions due to deforestation and drainage of peatlands.

Increased logging and deforestation are also taking place globally to support the bioenergy industry. The long term consequences of deforestation are scientifically acknowledged. Forests are responsible for consuming 25% of the CO_2 humans produce.

Agricultural biomass, competing with food production for arable land, is an increasing concern as well. To calculate land use requirements for different kinds of power production, it is essential to know the relevant area-specific power densities. Smil estimates that the average area-specific power densities for biofuels, wind, hydro, and solar power production are 0.30 W/m^2, 1 W/m^2, 3 W/m^2, and 5 W/m^2, respectively (power in the form of heat for biofuels, and electricity for wind, hydro and solar). The average human power consumption on ice-free land is 0.125 W/m^2 (heat and electricity combined), although rising to 20 W/m^2 in urban and industrial areas. The low yields of biofuel make it unattractive compared to other renewables.

Conclusion

Building a renewable energy future

Technological advancements, government policies and consumer awareness have played a significant role in advancing the renewable energy cause in the United States and worldwide over the past decades. For example, the solar Investment Tax Credit (ITC) has significantly contributed to solar energy growth in the United States. ITC is a federal incentive program passed by congress in 2006 to support solar energy development. It provided tax credits to residential and commercial customers who installed solar heating/cooling, concentrating solar and photovoltaic solar technologies. The program has also helped usher in billions of dollars in investments and job creations to local economies. On a localized level, many states, counties, municipalities and utilities designed policies and offered incentives to develop their renewable energy industries further. Among the most aggressive state policies is California's recent announcement to end all new gasoline vehicle sales in the state by 2035.

Federal subsidies for wind and solar projects and technology development totaled about $75 billion over the past decade, according to data from the Treasury Department, Congress' Joint Committee on Taxation and the Congressional Research Service. The Solar Energy Industries Association attributes the investment tax credit in the past ten years to growing the industry by more than 10,000 percent and leveraging $140 billion in economic activity. The solar industry now employs more than 340,000 people and is the fastest-growing job market in the U.S.
Government sources, such as The Database of State Incentives for Renewables & Efficiency (DSIRE), also played a significant role in consumer awareness. The database provides a comprehensive list of solar incentives and policies for each state. Additional government programs, such as the U.S. "Energy Star" program, have focused on educating the public on the benefits of energy-efficient products. The energy efficiency sector has saved and continues to saves Americans $billions in utility bills. The industry also creates more jobs than the fossil fuels industry.

The tax credits for wind and solar energy have also contributed to a remarkable drop in the cost of renewables projects. In the past decade, wind energy cost have declined nearly 70 percent, while utility-scale solar prices fell by almost 90 percent. Solar and wind power are now cheaper sources of energy than coal, nuclear and natural gas technologies. For the first time, renewable energy sources can be evaluated on an equal footing with traditional fossil fuels based energy. In addition, there are other favorable policies and possibly incentives on the horizon. The Democratic Party's victory in the U.S. November 2020 election is expected to sprint forward a much-debated plan, "The Green New Deal" to the forefront. According to party officials, it is a program designed to rebuild the U.S. economy on the backs of renewable energy.

Considering every corner of the globe is blessed with one form of renewable energy or another, local governments and communities have the freedom to make energy decisions independently. They are able to implement policies independent of other fossil fuel-rich regions and independent of fuel price fluctuations. Economic gains, such as investments and job creation, also remain local, benefiting local communities. Renewable energy also helps create a cleaner environment and improved public health indiscriminately. Therefore, a gradual shift from fossil-based energy to renewable energy is economically, socially and politically beneficial.

To those who believe the climate is changing for the worse, there are cleaner and renewable energy sources to promote. To those who deny climate change, if we are wrong about climate change, the worst we have done is cleaned up our environment and developed additional energy sources for our consumption. It's a win-win for all.

References

A new translation by Robin Waterfield, "Herodotus, The Histories" Oxford world's classic, 1998

American Clean Power Association (www.cleanpower.org)

American Oil & Gas Historical Society (www.aoghs.org)

Beloved Community Initiative (www.becomingbelovedcommunity.org)

Britannica (www.britannica.com)

British Library (www.bl.uk/georgian-britain)

Bureau of Ocean Energy Management (BOEM) (https://www.boem.gov)

Centers for Disease Control and Prevention (www.cdc.gov)

Dakota Access Pipeline (www.daplpipelinefacts.com)

DRC (https://peacekeeping.un.org/en/mission/monusco)

Earth Watch (www.earth.esa.int)

Earth works (www.earthworks.org)

Eco home (www.ecohome.net)

Energy sage (https://www.energysage.com/solar/101/types-solar-panels/)

Energy Start (www.energystar.gov)

Environmental and Energy Study Institute (EESI) (www.eesi.org)

Eritrean center for strategic studies (http://www.ecss-online.com/)

Explore the World of Piping (www.wermac.org)

Food and Agriculture Organization of the United Nations (www.fao.org)

Geo world map https://energyeducation.ca/encyclopedia/Geothermal_district_heating

History of Wind Energy in Encyclopedia of Energy Vol. 6, page 426

https://energyeducation.ca/encyclopedia/Pump_jack

https://oilprice.com/Energy/

Global Policy forum (https://www.globalpolicy.org/security-council/dark-side-of-natural-resources/timber-in-conflict.html)

Greenpeace (https://www.greenpeace.org/usa/arctic/issues/oil-drilling/)

International Solar Energy Sociel (ISES) (www.ises.org)

King Leopold's Ghost, by Adam Hochschild, copyright 1998 by Adam Hochschild; prologue, 18

National Energy Education Development Project (NEED) (www.need.org)

National Oceanic and Atmospheric Administration (www.noaa.gov)

National Park Service-U.S. Department of the Interior (www.nps.gov)

National Parks Conservation Association (www.npca.org)

National Renewable Energy Laboratory (NREL) (www.nrel.gov)

Natural Resources Defense Council (www.nrdc.org)

New Living Translation (www.bible.com)

New World Encyclopedia (www.newworldencyclopedia.org)

Radio Free Europe Radio Liberty (www.rferl.org)

Solar Energy Industries Association (www.seia.org)

The Travels of Marco Polo (https://en.wikisource.org/wiki/The_Travels_of_Marco_Polo/Book_2/Chapter_30)

U.S. Army Corps of Engineers (www.usace.army.mil)

U.S. Bureau of Labor Statistics (BLS) (www.bls.gov)

U.S. Bureau of Reclamation (https://www.usbr.gov)

U.S. Department of Energy (www.energy.gov)

U.S. Department of Labor (www.dol.gov)

U.S. Department of Transportation (www.phmsa.dot.gov)

U.S. Energy and Employment Report (USEER) (www.usenergyjobs.org)

U.S. Energy Information Administration (www.eia.gov)

U.S. Fish & Wildlife Service (www.fws.gov)

UN Environmental Programme (www.unep.org)

United Nations (www.un.org)

United Nations and the Rule of Law (www.un.org)

United States Environmental Protection Agency (www.epa.gov)

United States Geological Survey (www.usgs.gov)

Wikipedia (https://en.wikipedia.org/wiki/Hydraulic_fracturing)

Wikipedia (https://en.wikipedia.org/wiki/Thermal_mass)

World Health Organization (www.who.int)

Photo credits

The majority of the photos and illustrations in this book are taken by the author. The photos listed below are courtesy of various individuals and agencies.

Vertical Blades: Nashtifan, Iran: https://www.irandestination.com/asbads-of-iran/ (pg. 11)

Waterwheel: Paul Brennan from Pixabay: grist-mill-1572660_1280 (pg. 13)

Hoover Dam: RJA1988 from Pixabay: hoover-dam-3780254_1280 (pg. 13)

Mosaic: Tuna Ölger from Pixabay: mosaic-1927649_1280 (pg. 15)

IOT Image: Gerd Altmann from Pixabay: city-3213676_1920 (pg. 20)

Power plant: Pixabay power-plant-1892407_1920 (pg. 22)

Yantai City, China: Götz Friedrich from Pixabay: china-3993149_1920 (pg. 23)

Mountain Strip Mining: Image by Free-Photos from Pixabay: pit 984037_1920 (pg. 24)

Strip Mining Acid Lake: Strip Mining Photo by Dion Beetson on Unsplash dion beetson oF7hh97lVqA unsplash (pg. 24)

Underground Coal Mining: Photo by Arno van Rensburg on Unsplash_arno-van-rensburg-35uadY2c0DE-unsplash (pg. 24)

Pump jack: Image by mhouge from Pixabay: pumpjacks 3804889_1920 (pg. 25)

Oil/tar sands:

Water pollution: jwvein from Pixabay: monolithic part of the waters 3137978_1920 (pg. 27)

Green toxic pond: Image by enriquelopezgarre from Pixabay_mines-4944647_1920 (pg. 28)

Toxic ground: Image by Dimitris Vetsikas from Pixabay mine 3811241_1920 (pg. 28)

Theodore Roosevelt National Park, North Dakota, USA: Photo by Intricate Explorer on Unsplash_intricate-explorer-2MIPToQmVY4-unsplash (pg. 29)

Arctic nights: Sturrax from Pixabay aurora 2232730_1920 (pg. 34)

Arctic nights: Noel Bauza from Pixabay_aurora_1190254_1920 (pg. 34)

Image by Lisa Redfern from Pixabay_mesa-verde-709721_1920 (pg. 51)

Ireland home: Image by David Mark from Pixabay_ireland-69817_1920-1 Energy Star (pg. 54)

Energy Star pics: https://www.energystar.gov/ (pg. 61)

Eritrean dams: Eritrean Center for Strategic Studies (pg. 69, 77)

Lake Mead: National Park Service: U.S. Department of the Interior (pg. 70)

Hoover Dam Image by MrUweS from Pixabay Arizona 830472_1920 (pg. 70)

Hoover Dam aerial view Image by Regina Shanklin from Pixabay hoover dam 2038951_1920 cut (pg. 70)

Chief Joseph Dam: U.S. Army Corps of Engineers (pg. 78)

Pelton turbine: Wolfram Linden from Pixabay_metal-3707932_1920 (pg. 81)

Kaplan diagram: USGS.gov (pg. 83)

Steel tower: (Joenomias) Menno de Jong from Pixabaywind-turbine-3840422_1920_mod (pg. 93)

Lattice tower: Lolame from Pixabay_wind-power-5457775_1920 (pg. 93)

Composite wind blade parts: German wind energy association (pg. 94)

Blade transportation: Bishnu Sarangi from Pixabay_long-vehicle-320296_1920 (pg. 95)

Nacelle left: 3444753 from Pixabay_wind-turbine-1735776_1920 (pg. 97)

Nacelle center: energy-4650105_1920 from Pixabay (pg. 97)

Nacelle right: Steffen Horstmann from Pixabay_pinwheel-642532_1920 (pg. 97)

Inside of wind turbine/Nacelle: U.S. Department of Energy (pg. 98)

Installation left: Erich Westendarp from Pixabay_wind-power-3045169_1920 (pg. 102)

Installation center: Hans Linde from Pixabay_windrader-2759645_1920 (pg. 102)

Installation right: Hans Linde from Pixabay_windrader-2759650_1920 (pg. 102)

Onshore wind farm: David Mark from Pixab_wind-farm-1771238_1920 (pg. 104)

Offshore wind farm: Anette Bjerg from Pixabay_turbine-till-3058419_1920 (pg. 106)

Horizontal (HAWT): (Joenomias) Menno de Jong from Pixabay_eco-friendly-2232415_1920 (pg. 107)

Vertical (VAWT): Rafael Albaladejo from Pixabay_energy-4734634_1920 (pg. 107)

Solar panel details: Based on pic from www.solarchoice.net.au (pg. 114)

Solar panel: Sebastian Ganso from Pixabay_photovoltaic-system-2742302_1920 (pg. 114)

Bifacial solar panel: National Renewable Energy Laboratory (NREL) (pg. 116)

Periodic Table: ExplorersInternational from Pixabay_science-2227606_1920 (pg. 117)

Solar panel system connection diagram: (pg. 118)

Fresnel-Collector: U.S. Department of Energy: Ferrostaal-Hauke-Dressler (pg. 122)

Parabolic: U.S. Department of Energy (pg. 122)

Ivanpah Solar plant _ USGS_USDOE (pg. 124)

Yellowstone: pixabay image/Darilon_nature-3275686_1920 (pg. 125)

Plant power: pixabay image_power-plant-67538_1920-1 (pg. 133)

Danakil region: Afrikit from Pixabay_ethiopia-634222_1920 (pg. 137)